Walter de Henley, Elizabeth Lamond

Walter of Henley's Husbandry,

Together with an Anonymous Husbandry, Seneschaucie, and Robert Grosseteste's

Rules

Walter de Henley, Elizabeth Lamond

Walter of Henley's Husbandry,
Together with an Anonymous Husbandry, Seneschaucie, and Robert Grosseteste's Rules

ISBN/EAN: 9783337167349

Printed in Europe, USA, Canada, Australia, Japan

Cover: Foto ©berggeist007 / pixelio.de

More available books at **www.hansebooks.com**

WALTER OF HENLEY'S
HUSBANDRY

TOGETHER WITH

AN ANONYMOUS HUSBANDRY, SENESCHAUCIE
AND ROBERT GROSSETESTE'S RULES

THE TRANSCRIPTS, TRANSLATIONS, AND GLOSSARY
BY
ELIZABETH LAMOND, F.R.Hist.S.

WITH AN INTRODUCTION
BY
W. CUNNINGHAM, D.D., F.R.Hist.S.

LONDON
LONGMANS, GREEN, AND CO.
AND NEW YORK: 15 EAST 16th STREET
1890

All rights reserved

PREFATORY NOTE

THE volume which is now offered to the public was undertaken in consequence of the interest which was aroused by a paper read before the Royal Historical Society in May 1889. It had been previously intended to print Walter of Henley's *Husbandry* in the Appendix of the *Growth of English Industry and Commerce during the Early and Middle Ages* (Camb. Univ. Press, 1890); but the author gladly gave way when it became apparent that the Council of the Society would arrange for the publication of the four treatises which are so intimately connected. The task of selecting the best text for transcription and of translating has involved much more work than was apparent at first; where difficulties are still unsolved they have been stated in a manner which may assist other students in the endeavour to carry the investigation further. Dr. Braunholtz, Lecturer in French in the University of Cambridge, has kindly read the proofs and made many valuable suggestions, and we take this opportunity of expressing our in-

debtedness to him for the constant assistance he has rendered. We also desire to thank the authorities of the Cathedral Library at Canterbury, and of the Heralds' College, for access to copies of the different treatises in their charge, and the Master and Fellows of S. John's College, Cambridge, and of Merton College, Oxford, for allowing their MSS. to be transcribed.

E. L.
W. C.

August 1890.

CONTENTS.

	PAGE
PREFATORY NOTE .	iii
INTRODUCTION . .	vii
WALTER OF HENLEY'S HUSBANDRY . .	1
TRANSLATION OF WALTER OF HENLEY'S HUSBANDRY, ATTRIBUTED TO ROBERT GROSSETESTE	89
ANONYMOUS HUSBANDRY	59
SENESCHAUCIE .	83
GROSSETESTE'S RULES .	121
GLOSSARIAL INDEX	151

INTRODUCTION

THE four treatises which are brought together in this volume are of interest from many points of view. A work attributed to Robert Grosseteste deserves to be rescued from oblivion, and the peculiarities of the dialect in which these writings were composed may attract the attention of students of early French. Others may, however, be led to examine them from an interest in the subject-matter of which they treat, and this has been the primary consideration in planning the present edition. The work of an author who is only known through this book, and two anonymous treatises, have been placed alongside the maxims which were laid down by the great Bishop of Lincoln, for they deal, and deal at greater length, with the management of estates. Public attention has been called to the treatise of Walter of Henley by Professor Thorold Rogers, who has made frequent reference to it in the earlier volumes of his great work on *Agriculture and Prices*, and it is most desirable that the first-hand evidence on this important side of the life of our forefathers should be rendered generally accessible to students of English History. This is more especially the case, as the various treatises, in so far as they have been printed already, have been curiously mangled; it has been an interesting task to try and disentangle the confusion into which they have been thrown, and present them as nearly as possible in their original form, while at the same time the whole of the interpolated matter has also been retained. The student has now access to

the thirteenth century treatises, and the means of noticing the later accretions.

At the same time, while the main object of the Royal Historical Society in undertaking this work must naturally be to render the whole of the matter these treatises contain accessible to the members, the literary interest which attaches to these writings has not been altogether neglected. It is hoped that the work of grouping the MSS. and discriminating the insertions has been so far accomplished as to facilitate the work of anyone who may hereafter undertake the work of issuing a critical edition of the text. Several of the fourteenth century MSS. are written in French closely corresponding to that in the Statute Book, but the earlier MSS. give readings which appear to be somewhat more colloquial; and there is very great difficulty in attempting to reconstruct the language that Walter of Henley personally used, either in dictating or in writing. This difficult task has not been attempted in the present edition; the text in each case is a simple transcript of a single MS., which has been reproduced, without correction, even in those cases where there were obvious slips on the part of the transcribers. Though it was tempting to make a more ambitious effort, there is reason to believe that, by supplying a text which has not been tampered with under pretence of correction, the Society has furnished a sound basis to students for further investigations.

In this translation, also, the requirements of the modern historical student have been kept in view, as the main object has been to render the matter of these treatises readily accessible. In no case, therefore, has the English been sacrificed in the attempt to reproduce the construction of the French sentences as closely as possible, while the text affords the means of verifying the substantial correctness of the translation. Doubtful and disputed renderings are for the most part discussed in the Glossary.

The treatises are distinctly practical, and were intended to assist men in the ordinary business of life; they assumed on the part of the readers a familiar knowledge of institu-

tions and practices that have long since passed away. Though they so far supplement one another as to give a fairly complete picture of the life in a thirteenth century manor, it may be convenient to introduce a brief sketch of the organisation of a mediæval estate, and to trace the causes which brought about the destruction of the system that was then in vogue.

This introduction will deal in turn with the following topics:—
    I. Thirteenth century estate management.
    II. The relations of these treatises.
    III. Walter of Henley's *Husbandry*.
    IV. The anonymous *Husbandry*.
    V. *Seneschaucie*.
    VI. Grossoteste's *Rules*.

I.

THIRTEENTH CENTURY ESTATE MANAGEMENT.

There are many contrasts between the landed system which obtains in this country at present and the mode of estate management which was in vogue when Walter of Henley wrote. Perhaps the two most important differences are these: (1) That there are now three classes concerned in agriculture where formerly there were only two; and (2) that the economic obligations of these classes toward each other are now discharged in cash, whereas in former days they were often paid either wholly or in part in kind.

1. We habitually classify the agricultural population into the landlord who owns the land and buildings and receives rent, the farmer who supplies the capital and superintends and looks for profit, and the labourer who does the work and is paid a weekly wage. In all probability the landlord has very little, if any, land in his own hands, and though he may sink capital in permanent improvements, he does not feel called upon to provide the stock on the

farms. In the thirteenth century, however, there were only two economic classes,[1] not three, and the lord had a large *home farm* or *domain land* which was his chief source of income.

This home farm was cultivated by the labour of the villans, who were required to spend on it so many days each week, according to the season of the year, and who had to discharge certain occasional services as well. In return the lord supplied them with small holdings which they could cultivate on their own account on any days when they were not obliged to be at work on the domain lands; and the villan when he entered on his holding received an outfit from the lord of the stock which was requisite to work his holding. The villan may thus be regarded as a labourer on the home farm, whose regular days of labour (week work, *dies operabiles*) and times of special employment (boon work, *precariæ*) were accurately defined, and whose rations on his working days were carefully specified, but who had all the rest of his time to himself, and whose chief maintenance came from his own holding. The villan was not paid wages like the modern labourer, but in an ordinary way he worked three days a week on the domain land, he gave besides extra days in autumn, and performed other incidental duties, and he had in return his meals on some of the working days, and a holding (*virgata*) of about thirty acres, which his lord stocked with a yoke of oxen and half-a-dozen sheep when the villan first entered on his tenancy.

By a process which has many parallels in feudal times, the obligations which originally attached to the villans personally came to be connected with the holdings they enjoyed. We thus get the whole of the villan's *virgates* spoken of as servile land, and the obligations of work on

---

[1] There were, of course, an extraordinary number of social grades, whose status differed for legal purposes; the statement above has mere reference to an economic grouping. It also ignores the free tenants (and the *redditus assisæ*, cf. p. 60), as their relations to the manorial lord appear to have been fiscal, and they lay outside the ordinary processes of agriculture as he was concerned in directing them.

the domain land can be best regarded as incident to the tenure of a piece of servile land. We might thus have a freeman who was not personally of villan status, but who had undertaken to work a plot of servile land, and was to that extent obliged to render his quota of villan services on the domain land. From this point of view it is easy to see that the relation of the villan to his lord may be compared, not with that of the modern labourer, as is done above, but with that of the modern farmer. He may be regarded as a tenant who received a certain area of land ready stocked, and who in return paid a rent in the form of service.

The successful working of the domain land was the really important thing for the lord in considering the management of his estates; and the main element of success depended on the labour available. The science of agriculture was in its infancy, root crops were unknown, and there were no artificial grasses; a system of rotation as practised in the present day was impossible. On some estates each field was allowed to lie fallow every other year (two-field system), on other estates every third year (three-field system), and these different methods are described in Walter of Henley's treatise. Though they were accustomed to manure the land, and in some districts used marl, they did comparatively little in the way of drainage.[1] Modern methods of agriculture and the modern application of capital to land were impracticable. The one necessity was labour; from the estate which was well stocked with men and with oxen a fair income could be derived; but if there was no labour available, the estate could only have a prairie value. The consequence is that the variations in the value of land at any time of social convulsion might be very great; if the labourers deserted the estate, or died, the value declined immensely; while if the estate were restocked with labour, its value would be restored almost as rapidly as it had sunk. The curious variations in the value (not merely in the rating) of the

[1] See, however, Thorold Rogers, *Agriculture and Prices*, I. 19, and in the Glossary,' s.v. *asseuer*.

estates during the disturbance caused by the Norman Conquest, and still more after the Black Death, only become intelligible when we remember how closely the value of the estate depended on the maintenance of a sufficient head of labour.

This, then, was a fundamental principle which underlay the whole scheme of rural management, that the persons of the labourers should be retained on the estate, and that their progeny should not be permitted to avoid becoming liable to the same obligations in turn. The whole of the social restrictions on villans putting their children to school, or allowing them to be ordained, or to be apprenticed in a town, had this as their economic justification—that there must be available labour for the home farm. These restrictions were so universally known that Walter of Henley did not think it necessary to allude to them; he is rather concerned with the other question as to the means of getting their full quotas of work out of the servile tenants. And he points out important aid which may be secured in this matter by laying stress on the communal obligations of the villans. The villans as a whole were responsible, and if anyone failed to perform his service, the others might be required to make good his deficiency. It is obvious that this principle of communal responsibility might be urged so as to be very oppressive, and with the view of exacting far more than the defined obligations of the tenants. It offers, like so many of the matters touched upon, an interesting subject for investigation, as it would be curious to know how far it was successfully insisted upon, and whether this claim formed an element in fomenting the discontent which followed the Black Death.

To carry out this system of management it was of course necessary to have a considerable staff of officials. The bailiff was appointed by the lord to look after the whole estate in detail; his duties closely resembled the occupations of a modern farmer, as he was responsible for all the stock on the home farm, as well as for seeing that the labourers paid their proper services; he was directly

responsible to the lord for everything connected with the prosperity of the estate, and had to account in great detail for everything under his charge. The two minor but very important officials were the *prepositus* and the hayward (*messor*), who were somewhat like foremen labourers; the former seems to have been the official representative of the villans, who was responsible for them; the latter had to supply their contributions of seed, and to be present to superintend their work. A large landed proprietor, who had estates in different places, would also require a seneschal (or steward) who should represent the lord personally and hold the manorial courts on his behalf. His duties were legal rather than economic, though he had the general superintendence of everything on the estate; he was not, however, possessed of complete authority, as he could not dismiss a responsible servant, like a bailiff, without the consent of the lord himself.

2. The origin of the legal powers of the lord of the manor and the actual conditions which gave rise to the different types of villanage do not concern us here; it may suffice to say that the manor, as an economic institution, has interesting analogies in earlier history as well as in backward countries in the present day. Wherever we have an estate organised as a whole, and which has very little communication with the outside world, it will be difficult for the owner to realise the produce of the land or to make purchases. His object will be to render the estate, so far as may be, self-sufficient. He will consume or store the produce of the estate, and he will endeavour so to utilise its resources that there will be little need to purchase anything from the outside. This principle is plainly laid down in Charles the Great's Capitulary *de Villis*, it is explicitly stated in Grosseteste's *Rules*, and it may be taken as a fundamental maxim for the judicious management of land, as understood in days when roads were bad and the opportunities of trade infrequent, since the business of the country was mainly carried on at annual fairs. It follows almost necessarily that the landowner had comparatively

little to offer for sale; if his granary was full and all possible requirements were provided for, he would be willing to sell the surplus corn, if he could get a good price for it. But if the harvest had been a scanty one, he would have little if anything to spare, as he had to look to home requirements first of all, and if the price was very low he might prefer to store his grain rather than sell it at a very low figure. This is a statement of the policy he aimed at, but in practice it could not be fully carried out; metals and salt would be wanted on estates where they could only be procured by trade. It would hardly have been possible to levy a Danegeld unless rural proprietors generally were accustomed to sell a portion of the produce and to lay up a hoard of silver; but the circulating medium was not always available, and fiscal obligations were sometimes discharged in kind as late as the time of Henry I. The rural proprietors frequently had to buy and attempted to sell, but they had no occasion to open up new markets or develop internal trade, as the policy of each was to do without trade as far as possible.

The surplus available for sale would, generally speaking, bear a very small proportion to the total crop. The seed was generally provided by the villans, but the necessary stores for their rations on boon days, together with the supplies for the lord and his household, must have absorbed the greater part. In some estates, where the owner was not even an occasional visitor but an absentee, there would be no need to retain provisions for his table; hence on monastic estates, when situated at a distance from the monastery, it would be convenient to sell a larger proportion of the produce and to transmit the value to the abbey. It is thus obvious that, even on adjoining estates, there might be considerable differences in the degree in which their respective managers ventured to have recourse to trade; some might be able to sell half the produce or more in ordinary years, while others would be forced to retain almost the whole crop to meet the requirements of the villans and the household.

It is hardly conceivable that there could be regular

money payments within an estate so long as it was practically isolated and there was little regular trade. If the lord did not sell his corn, he could have no silver with which to pay labourers' wages; if the villan did not sell some of the produce of his holding, or earn a day's wages by working now and then in his own time for the lord, he could not buy his freedom from predial service by regular cash payments. But as the estates lost their isolation, and the habit of selling a large proportion of the produce became more common, the conditions were present in which the lord could begin to receive payments in lieu of service, and to hire labourers to work his home farm if he preferred this system. There was thus in the twelfth century a gradual approximation to more modern conditions on many estates; the home farm was worked by hired labourers who received wages; while the villans had bought themselves off from the obligation of doing the customary work by paying a quit-rent. The increase of the practice of selling produce off the land, and of satisfying the mutual obligations of the dwellers on the estate in cash, would go on together. Within the present century the railway system has opened up facilities for trade in dairy produce which did not previously exist, and allowances in milk are less frequent than formerly. But it would seem that the practice of commuting service for cash payments had begun to show itself in some places before the Conquest;[1] and it appears to have advanced steadily in all parts of the country as the conditions became present which rendered cash payments possible between the lord and the labourers or tenants.

Even if the obligations were rarely discharged in cash by the villans in the time of the Confessor, the relation between their duties and money, i.e. the cash equivalent of their services, could be stated with precision. In every Domesday entry in turn we find an estimate of the worth of the estate, which was, of course, mainly dependent on the villans' obligations, and this is stated in terms of money.

---

[1] *Domesday Book*, I. 314, a. 2. Hotun in Yorkshire.

It was, apparently, argued in later times, from the mode of statement adopted in the Great Survey,[1] that the predial services had been all commuted for money before the Norman Conquest. But the practice of estimating in terms of money the value of obligations which were discharged by actual service was common in the thirteenth century; and though it is clear that the process of commutation had already begun in the time of the Confessor, there is reason to believe that it was quite exceptional. A very ordinary arrangement in the time of Edward I. appears to have been that the villans' obligations were stated in terms of money, but were paid either in services or in cash at the will of the lord. If partly paid in cash, the villan would still have to defray his remaining duties and dues in service or kind. When the villan paid to be quit of the ordinary week work, and, so to speak, bought his time for himself during any one year, his payment was entered as *opera vendita*. Some were able, however, to pay cash every year and to be quit of these services quite regularly; they are sometimes spoken of as *molmen*;[2] or the fact that the lord had agreed to take cash in lieu of all services[3] is noted in the *Extent*, which describes the condition of his estate, or is drawn out in detail in a special document.[4] When this stage was reached, the villan, with his virgate of thirty acres in the common arable fields, was in a position very similar to that of the modern peasant farmer who pays rent for his land.

The change by which one portion and another of the villans' obligations were commuted for cash seems to have gone on slowly, here and there, all over the country. When the villan had the opportunity of paying cash he might possibly be glad to buy his freedom, and be his own master all through the week; this was certainly the case after the Black Death, when his labour had become much more valuable, as it would obviously be to his interest then

---

[1] 1 R. ii. c. 6.
[2] Vinogradoff in *The English Historical Review*, I. 734.
[3] *Hundred Rolls*, II. 636.
[4] *Growth of English Industry during the Early and Middle Ages*, p. 513, for the case at Barrington in Cambridgeshire.

to buy his own time from the lord at the old customary rates if he could. In the thirteenth century, however, it may have been the lord rather than the villan who preferred to insist on cash payment where he could, as there was less trouble in collecting the payments than in superintending the regular work day by day throughout the year; and there was less difficulty about the quality of the coin than about the diligence of the service. It thus appears that, from the side both of the villan and of the lord, the change from actual services to cash payment would be welcome, where the circumstances of the estate rendered it possible.

This change in the relations between the villan and the lord necessitated other changes in the management of the home farm. The lord might work his domain land with hired labour, or he might break it up into holdings and let it, or he might use it as a sheep farm; the last expedient was increasingly adopted after the Black Death. This terrible catastrophe gave a blow to the system of farming with hired labour, as the rates of wages became abnormally high, and it could scarcely be remunerative. The estates on which bailiff farming had most chance of surviving were those where there had been least change, and where the actual services were still due. Even before 1349 the lords on some estates had broken up the home farm and rented it in portions to the tenants, who were collectively responsible for the payment of their money-rents, as they had formerly been for rendering predial services. An instructive case occurs in the letting of the St. Albans' estate at Granborough in Buckinghamshire in 1346.

It is now possible to describe with greater precision the nature of the topics treated in each of the four treatises which are here printed. The general subject is the management of estates, and in the management of an estate the successful working of the home farm—under the superintendence of a bailiff, and by means of the services of the villans, with the assistance of hired labourers—was the chief element of success. They deal primarily with bailiff farm-

ing, and with bailiff farming as it was organised in the thirteenth century. All four treatises apparently date from this period, though only one of them can be assigned with precision. There must surely have been some remarkable conditions at that time which resulted in the production of these independent treatises, which so met the requirements of the English landlord that no serious attempt was made to supersede them till the sixteenth century.

Professor Thorold Rogers fixes on the reign of Henry III. as the time when the practice of keeping written accounts on each estate became general;[1] this was another symptom, and a very important symptom, of the increased care which was devoted to the management of estates. Possibly the general political conditions were favourable to rural prosperity, but it is also probable that the industrial and commercial activity of the twelfth century had begun to react on rural affairs. The Crusaders had given an impulse, under which foreign trade had flourished greatly, and many towns had grown into some importance. There may have been some direct influence in the planting of new monasteries and experience of new settlers, but at any rate the agriculturist would have new markets, and be tempted to realise his produce more readily; and as the separate estates were drawn more and more into the stream of internal commerce, the difficulty of superintending the management and checking the servants would increase, unless they were compelled to keep regular accounts. The greater responsibilities of the bailiff, who not only superintended the villans but sold a larger proportion of the produce off the estate, rendered it necessary that accounts should be regularly kept; and the mere fact that estate management had become more complicated and difficult sufficiently explains why it attracted more attention, and called forth these systematic treatises.

[1] *Agriculture and Prices*, I. 2.

II.
THE RELATION OF THESE TREATISES.

The four treatises which are included in this volume have many interesting points of connexion; they contain practical hints set forth by practical men to assist others in the management of their affairs; they deal simply with matters of ordinary experience. Nor do the authors attempt to follow and apply the principles of any classical authority; it is a genuine effort to put on record the unwritten wisdom of the time. Hence they represent a fresh and genuine literary effort of certain Englishmen who wrote about agriculture in a thoroughly English spirit. Fragments of English speech crop up here and there, and give a sufficient flavour of our soil, but the whole dialect is the Anglicised Norman French, of which few prose specimens survive outside the Statute Book. Offensive as it is to the eye of a French scholar, the language clearly indicates the insular origin of the treatises which employ it.

While these treatises are thus similar, and while they to some extent overlap, they also serve to supplement one another, as the precise object is different in each case. The treatise of Walter of Henley is a survey of each of the departments of rural economy in turn; ploughing and harrowing and other operations come within his view; he gives suggestions to enable the lord to avoid the leakage which occurs so easily when there is no careful supervision, and his treatise supposes that the lord would look into everything himself. It is rightly entitled *Husbandry*, not because it has to do with tillage, but because it shows how he may husband his resources and manage thriftily.

The anonymous *Husbandry* is primarily concerned with the estate accounts; it advises the traditional policy of rendering the estate as self-sufficient as possible, but it describes how the accounts should be kept and passed, while it gives rough estimates which may enable the lord to check the rates at which the bailiff calls on him to pay.

It has less reference to the actual management of the land itself than to the accounts of the bailiff and the means of checking them.

The *Seneschaucie* is even further removed from the details of rural employments; it deals with the duties of each officer in turn, and describes the functions of the steward, the bailiff, the præpositus, and so forth. It reveals a curious division of the labour of superintendence, and sets forth the relations of these officers to one another in some detail. It is more wordy and formal than the other writings, and has closer relations than theirs to legal documents, such as those contained in the Statute Book.

Grosseteste's *Rules*, though quite as practical as Walter of Henley's *Treatise*, were intended for a great Countess, who could not possibly look into everything herself. It brings in a side of life which the other treatises leave untouched, as it deals not only with production but consumption, and lays down maxims for the management of the household.

This last work appears to have had but little circulation; though it deals with many topics, the treatment is somewhat slight, and even the great reputation of the author did not prevent it from falling into oblivion. The *Seneschaucie* ceased to be of much interest as the services of the villans were commuted, and the personal presence of the hayward and others was no longer demanded. As bailiff farming gradually ceased to pay and disappeared, the instructions in the anonymous *Husbandry* for checking accounts were also out of date. The one treatise which had real vitality was that of Walter of Henley; his practical hints on the details of rural life had a lasting importance; it continued to be in frequent use and in wide circulation; additional matter was incorporated in it as the changed circumstances of agriculture demanded; it was thought worthy of re-issue from an English printing-press for the guidance of practical men, more than two hundred years after it was written.[1] At length it lost its position as the best book on the subject,

---

[1] See below, p. xxxix.

because it was superseded by the work of Sir Anthony Fitzherbert,¹ in which, however, much of the earlier treatise was incorporated without acknowledgment. Even then, however, it was not wholly forgotten; Gervase Markham, when writing on the *Enrichment of the Weald of Kent*, quotes Sir Walter of Henley as an authority who recommended that marl should not be ploughed into the soil, for while the virtue of dung will ascend, however deep it lie, 'marle sendeth his vertue downward, and must therefore be kept aloft and may not be buried in any wise,' p. 11. He also refers to 'Books of gainage or husbandry that were written in the days of Edward II. or before,' p. 4, a double mode of reference which we also find in Lambarde's note-book, mentioned below.

III.

WALTER OF HENLEY'S *HUSBANDRY*.

1. Of Walter of Henley, only one thing is known for certain. He had served the office of bailiff, for he makes incidental mention of the fact. The title of the manuscript (13) marked Dd. vii. 6 in the Cambridge University Library gives some further details, but on what authority is unknown. It runs as follows: 'Ceste ditee si fesoyt sire Waltier de hengleye qui primes fu chiualier e puis se rendesist frere precheur e le fist de housebonderie e de gaygnerie e de issue de estor.'

This gives us broad limits as to the date when the author flourished; the Dominicans came to England in 1221, and settled in Holborn and at Oxford during that year. It is most unlikely that the *Husbandry* was written before the thirteenth century opened; while, on the other hand, there is a MS. of the treatise at Canterbury written, as I am told by Dr. Sheppard—in the hand of John de Gare, who was clerk to the Prior—in the early years of Edward I. Within these limits there are no precise data available at present for fixing the exact time when the

¹ *On   bandry.*

treatise was written, though it is not impossible that some mention of this Walter may sooner or later turn up.

If a guess were hazarded as to the direction in which it would be best worth while to begin such a search, there are some grounds for examining the splendid collection of records in the Cathedral library at Canterbury. Walter of Henley's treatise was known at Canterbury at a very early date; it continued to be prized and copied at Christchurch, as no fewer than four of the existing manuscripts were written there. Curiously enough, there are two or three Canterbury manors in the immediate neighbourhood of Henley, at Monks Risborough, Newington, and Brightwell. It has been suggested to me as at least conceivable that Walter was a trusted retainer of the great abbey, who might be called on to serve as a man-at-arms, and who discharged the responsible duties of bailiff on these manors; and there may be documentary evidence which would favour this ingenious suggestion.

The argument which Professor Thorold Rogers bases on the silence of the treatise in regard to the scab does not seem very forcible.[1] Cases of this disease are not known to have occurred in England before 1283, and the silence of Walter of Henley shows that he wrote before that time, since he discusses another disease—the rot—so fully. But the passage which deals with the rot is an interpolation, which is found in one family of MSS. from the end of the thirteenth century onwards; and the transcribers would apparently have had no scruple about introducing directions for the treatment of the scab, if it had occurred to them to do so.

2. As has been stated above, the editions of Walter of Henley contain his treatise in very mutilated forms, and the manuscripts present a curious variety both in the matter they contain and in the arrangement they adopt. Some have considerable insertions; but it does not seem altogether impossible to arrange the MSS. in groups, and to indicate briefly the relations of these various groups to

[1] *Agriculture and Prices*, I. 460.

one another. The clearest principle for discriminating the different families appears to be found by noticing the divisions and headings of the chapters, and this will serve to distinguish four well-marked groups of MSS.

It may be convenient, however, to give a list of the MSS. which have come under my notice, arranged in chronological order; the dates assigned can only be regarded as approximate.

1. Cambridge University Library, Ee. i. 1 f. 251, Edward I.
2. Canterbury, Register J, Edward I.
3. Heralds' College, Arundel MS. xiv. Edward I.
4. Paris, Bibliothèque Nationale, 7011 f. 57, Edward I.
5. Guildhall, Liber Horn, Edward II.
6. British Museum, Cottonian MSS., Edward II.
7. Merton College, MS. cccxxi. f. 153, Edward II.
8. Bodleian Library, Douce MSS. 98 f. 187b, Edward II.
9. Cambridge University Library, MS. Dd. ix. 38 f. 252, Edward III.
10. Canterbury, Register P, Edward III.
11. Cambridge University Library, MS. Hh. iii. 11 f. 77b, Edward III.
12. British Museum, Add. MSS. 6159 f. 220b, Edward III.
13. Cambridge University Library, MS. Dd. vii. 6 f. 52b, Edward III.
14. British Museum, Lansdowne MSS. 1176 f. 181, Edward III.
15. Trinity College, Cambridge, O, 9, 26 f. 98, Edward III.
16. British Museum, Harleian MSS. 493 f. 408, Edward III.
17. Bodleian Library, Digby MSS. 147 f. 1, Edward III.
18. Cambridge University Library, MS. Dd. vii. 14 f. 228, Henry IV.
19. British Museum, Sloane 686 f. 1, Edward IV.
20. British Museum, Add. MSS. 20,709, Elizabeth.
21. Bodleian Library, Rawlinson MSS. B 471 f. 16, Elizabeth.

A. The text of Walter of Henley appears to be preserved in its earliest arrangement, though with modernised spelling, in a fifteenth century MS. (18) in the Library of the University of Cambridge. The whole runs on without any further divisions into chapters than by the mark of a paragraph which occurs so as to mark sixteen separate parts. There are similar divisions in the copy (14) among the Lansdowne MSS. in the British Museum; this dates from the time of Edward III., and the commencement of each chapter is marked by an elaborate initial letter, but there are neither headings nor numbers. An earlier representative of this group is the manuscript (3) in the Heralds' College, which dates from the time of Edward I. But the

transcriber inserted headings to the chapters, and it appears from the table of contents, as well as from a long insertion on sheep farming, that he also made use of a manuscript belonging to the *Liber Horn* family. There is a second Cambridge University MS. (18) in which the same divisions occur, but the chapters are numbered though they have no headings. A third Cambridge University MS. (9), written at Reading, has the same arrangement, but the chapters are in some cases subdivided and headings are added throughout. It appears to belong to this group, but there are some differences in the text.[1] A second MS. also written at Reading (16) has been preserved in the British Museum, and is closely similar in all its features. In the whole of this group the division of chapters differs from that in the text now printed; thus the sentence about the team of oxen and horses (p. 10) begins a new chapter, as does that about fallowing in April on page 12; so, too, the chapter about preparing manure (p. 18) begins at the sentence which comes second in the text; the introduction gives only one of the English proverbs. This may be distinguished as the Reading family.

B. The second family consists of two MSS., and this arrangement of the text has been adopted in the present edition. The earlier chapters are clearly divided, and begin with a new line; but they have no titles, while the later chapters have. The long title of the treatise is also distinctive of this group, and the phrase, *gaynage de tere*, which occurs in it, is suggestive of the title under which Matthew Parker's MS. was known, and may indicate that it belonged to this family. One of the two extant copies is in the Luffield book at Cambridge (1), and the other (8) among the Douce MSS. at Oxford. The Luffield appears to be the earliest example now existing, with the exception of one of the Canterbury MSS. I suspect that the author of *Fleta* also used a MS. of this group.

[1] The calculation as to the ploughing required in the two-field and the three-field is omitted, and the chapter on cattle is differently arranged.

C. The Canterbury MSS. form a group by themselves; they have the same divisions as the Luffield group, but titles are given to all the chapters. There are some peculiarities which distinguish the text of this family from that adopted in all the other MSS. In the chapter on ploughing, the quarentina is described, by a slip, as forty perches[1] broad as well as forty perches long. The breadth is correctly given as 66 feet, however; but it is said that forty turns of the plough can be made in the acre, which would give eighty furlongs; and these eighty furlongs are stated to form four leagues of twenty furlongs. This league of two and a half miles is shorter than our present league, but much longer than the league mentioned in all the other MSS. This consisted of twelve furlongs only, or a mile and a half, and, when thirty-six turns of the plough were taken in the acre, the team travelled six leagues. Besides this difference, the Canterbury MSS. transpose the sentences about the weeks in the year, and the length of the furrows. In the chapter about the dairy there is another transposition: the information about cows runs straight on, and the sentences about ewes conclude the chapter. The transcriber had also omitted the first sentence of the chapter about preparing manure (*Vostre estuble*, p. 18), but added it at the conclusion of the whole treatise, where it is not particularly appropriate. This group consists of five MSS., four of which were copied at Canterbury, two (2, 10) are still in the Cathedral Library, one (12) is in the British Museum, and one (15) in Trinity College, Cambridge; and there is also a fourteenth century transcript (11) in the University Library at Cambridge, which gives no clear evidence of the place where it was written.

D. The remaining group consists of MSS. which adopt the arrangement used in the *Liber Horn*. This has the same divisions as the Luffield and Canterbury groups, but the name of Walter of Henley does not occur in the title of the treatise, and the separate chapters have lengthy

[1] But this is correctly given in the transcript in the Cambridge University Library, marked Hh. iii. 11.

titles. In the *Liber Horn* (5) these are prefixed as a table of contents to the whole, and also occur at the commencement of the separate chapters. The MS. (7) in Merton College Library is very similar, as it has not only the headings of the chapters, and the table of contents, but it also agrees with the *Liber Horn* in giving some additional matter. In the chapter on sheep farming there is a lengthy insertion on some of the diseases of sheep, which is still further amplified in the translations. This long insertion, together with the headings of the chapters, are so characteristic, that there can be little doubt about these MSS. forming one family, though there are differences, not only in the spelling, but also in the construction of the sentences. These would be perfectly intelligible, however, if the copies were made by dictation. The Paris MS. (4), from which Lacour printed his *Traité inédit d'économie rurale*, also belongs to this family, and several of the headings of the chapters have been incorporated with the text, while a new heading is added as well; as, for example, in Chapter ix. of Lacour. In this Paris MS., however, the whole matter has been rearranged, and the anonymous *Husbandry* has been incorporated with Walter of Henley's work, so as to produce a curious confusion.

As stated above, the scribe who wrote the Heralds' College MS. (8) had access to some representative of this group, and it appears to be the foundation of the translation into Latin, which occurs in a book (17), formerly belonging to S. Mary's at Merton in Surrey, and now among the Digby MSS. in the Bodleian. The scribe who wrote the Merton College MS. had begun by translating the first chapter of the anonymous *Husbandry* into Latin, but he appears to have tired of the task. The translation among the Digby MSS., though it is not exactly based on any of the French MSS. I have seen, appears to be connected with this form of the text, either directly or through the Heralds' College MS.; it contains Walter of Henley's name in the title, the insertion on sheep farming is still further expanded, and the whole treatise is divided in a

somewhat pedantic fashion. The chapter about the survey ends[1] with the sentence about the breadth of an acre being sixty-six feet, and a new chapter, *de modo arandi*, begins somewhat awkwardly with the next sentence (p. 8, line 12). It is not without interest to notice that the monastery which possessed this copy was one that had long had an important trade in wool.[2]

This group of MSS. with the long insertion also formed the basis of an English translation (19) which occurs in a late fifteenth century hand among the Sloane MSS. in the British Museum.

3. So far for the arrangement and divisions of the text; a few words may be added about the different MSS., and in so doing it will be convenient to deal with them not in their chronological order, but with reference to the libraries where they are now to be found.

CAMBRIDGE appears to be specially rich in this department. There are no fewer than five copies in the University Library, besides one at Trinity.

University Library.

(13) Dd. vii. 6 f. 52*b*. This is a large folio volume on parchment, containing a collection of statutes, together with the treatises of Bracton and Hengham. It apparently dates from the earlier part of the fourteenth century. It is interesting on account of the title, quoted above. In judging of the value of the information it affords regarding the author, we must remember that the text from which this MS. was copied must have been of an early type, since the chapters had no headings. It is divided into sixteen chapters, similar to those in the text, and distinguished by numbers, and the introduction omits the second of the two English proverbs. The last sentence of the chapter on inspecting cattle in the present edition (p. 22) is transposed in this MS., and stands at the beginning of the chapter, but in a slightly longer form.

[1] As in the Heralds' College MS.
[2] *Growth of English Industry during the Early and Middle Ages*, 549.

The most important difference from the text as printed is that this MS. contains a passage on the respective cost of ploughing on the two field and three field systems. It is an interesting sentence, and has been printed as note 1 on p. 8. From this it appears that when the two field system was in use the field under crop was partly used for wheat or rye, and partly for barley or oats; not entirely devoted to wheat or rye, as one might have supposed. It also shows that, although a much larger area was under crop in each year when the three field system was used, the expense of ploughing was the same on each system. If the land was laid out in two fields of 80 acres each, there would be 40 acres to plough before the wheat was sown, 40 more before the barley was sown, and 80 to be ploughed twice over in June, when the stubble of the second field was broken up and it was left fallow, i.e. $40 + 40 + (80 \times 2) = 240$. If the three field system were used, there would be 60 acres to plough before the wheat sowing, 60 acres to plough before the barley sowing, and 60 acres to plough twice over when the stubble was broken up in June, i.e. $60 + 60 + (60 \times 2) = 240$. This sentence occurs in most of the MSS. of the Reading group, but not in the two copies written at Reading itself; it is also found in the *Liber Horn*, and in the Canterbury MSS., but it does not occur in the Heralds' College MS. It may perhaps be suggested that it occurred in the original text, but that the sentence was omitted by the copyists, for whom it had ceased to have a practical interest. In so far as the three field system had been thoroughly established on any set of estates, this comparison of the cost of the two modes of cultivation would be of little importance, and it may have been omitted on this account. This suggestion might possibly be confirmed or disproved by an examination of the evidence as to the manner in which the Canterbury, Luffield, and Reading estates were cultivated in the fourteenth century, if the necessary evidence is available.

(18) Dd. vii. 14 f. 228. This is a parchment folio written in the fourteenth century, and containing Bracton,

Hengham, a collection of statutes, &c. The treatise of Walter of Henley has been inserted in the time of Henry IV. It is commenced on a blank leaf, and continued along the foot of eleven subsequent pages. It has no numbers to the chapters, and no headings; the text corresponds very closely with that of 13, but it is somewhat later in date, and is much more carefully written. This is one of the latest transcripts, but it appears to be of great value for settling the form of the original work, though the language and spelling have been modernised by the transcriber.

(9) Dd. ix. 38 f. 252. This is a parchment folio written by various hands in the fourteenth century; it belonged to the monastery at Reading. It contains, besides a collection of statutes, much interesting information about the relations of the abbey and the town. Some of this has been printed, from other sources, by Coates in his *History of Reading*, App. No. V. It is one of two copies of the *Husbandry* which were certainly written at Reading, and which I have taken as representing one group of MSS. The text is, however, not precisely similar to that in the other numbers of the group; the chief difference is the omission of the sentence on the relative cost of working land on the two field and three field systems.

(1) Ee. i. 1 f. 251. This is the MS. from which the text of the present edition is printed. It has been reproduced exactly, without any emendations, in accordance with the reasons stated above. The volume is a folio on parchment, and contains Bracton, Hengham, statutes of the realm, and documents relating to lands belonging to the Prior of Luffield. A considerable portion of the volume is of the time of Henry III., but the transcript of Walter of Henley appears not to be of quite such an early date; other treatises and documents have been added and inserted in the early part of the fourteenth century.

(11) Hh. iii. 11 f. 77*b*. This is a folio on parchment containing a collection of statutes, and it formerly belonged to F. Tate, a Reader in the Middle Temple in the time of

James I. A considerable interest attaches to this volume, as it contains three of the four treatises now printed. Considering how much they have been combined in early MSS., like those printed in *Fleta* and by Lacour (*vide sub* pp. xxxii, xxxvii), there is a satisfaction in finding them here side by side as independent treatises. Walter of Henley's treatise is in a handwriting of the fourteenth century, but that in which the other two treatises are written is somewhat later, and so cramped as to be very difficult to read.

The text and headings are those of the Canterbury MSS., but the transcriber has corrected at least one slip that runs through the other MSS. of this group. He gives the measurement of the acre as forty perches in length and four in breadth.

(15) Trinity College: O 9, 26 f. 98. This is a late fourteenth century transcript of the Canterbury family, clearly but not very carefully written, and without any special features of interest. It consists principally of statutes, royal charters, and some forms for leases, &c. It also contains a glossary, a few recipes, and a vision of S. Thomas of Canterbury in a later hand.

LONDON.

(12) British Museum. Add. 6159 f. 220 b. This is a fourteenth century MS. full of documents of various kinds relating to Christchurch, Canterbury; it is clearly written, and has been used to correct the text of the present edition where it was specially difficult to construe, and it seemed to be desirable to have recourse to another form of the text.

(14) Lansdowne 1176 f. 131. This is a fourteenth century MS. fairly written on vellum, but not in good preservation. The volume also contains Bracton and several statutes. The transcript of Walter of Henley is similar to that of Dd. vii. 14 in the Cambridge University Library. It may perhaps be regarded as the earliest representative of that family, for the Heralds' College MS. appears to combine two sources; like the others of the group, it has no numbers to the chapters and no headings.

(19) Sloane 686 f. 1. This is an English translation

written on paper in a late fifteenth century hand; the writing is not good, there are several careless omissions of words, and no attempt at punctuation; it presents many curious features, and has been printed below. It appears to be founded on the same French text as the Latin translation in the Bodleian at Oxford (see below p. xxxv), for it contains considerable insertions about sheep farming. It also contains a curious passage about gleaners pocketing corn, which is not found in any of the MSS., but which appears in *Fleta* II. 82 § 2, and in *Seneschaucie* (see p. 98). A statement of the English measures is also inserted as Chapter II. Besides these insertions, the text has also suffered from mutilations; the introductory chapter in particular is much curtailed. The title states that the treatise was written in French and translated into English by Robert Grosseteste; this is not impossible, perhaps, but it is surprising that Walter of Henley's treatise had been so much corrupted before the bishop's death in 1253. Though this, as we shall see below, is not impossible, it appears more probable that some knowledge survived of the fact that Grossteste had written *Rules* for the management of an estate and household, and that the transcriber supposed this treatise to be identical with that of the Bishop of Lincoln.

(6) The MS. of *Fleta* in the Cottonian Collection, from which Selden printed his edition, contains very large extracts from Walter of Henley's treatise. They occur in the final chapters of the second book. The framework of this part of the work is furnished by the *Seneschaucie*, but this is the merest skeleton, and the largest portion of the matter is taken not from the *Seneschaucie* but from Walter of Henley. The translator showed considerable skill in the use he made of his materials; the following table which refers to the chapters and sections in Selden's edition may be of interest as showing how the two treatises have been inter-combined. He also made use of other materials which were drawn upon by a later interpolator of Walter of Henley, but I do not know whence they were all derived: the *Extenta Manerii* of the Statute Book is included.

C. 71.	§ 1.	Seneschaucie.
	§§ 2, 3.	Walter of Henley.
	§§ 4–17.	Extenta Manerii in the *Statutes of the Realm*.
C. 72.	§ 1.	Seneschaucie.
	§ 2.	Walter of Henley.
	§ 3.	Unidentified.
	§§ 4, 5, 8.	Walter of Henley.
	§§ 6, 7, 9–21.	Extenta Manerii.
C. 73.	§ 1.	Seneschaucie.
	§§ 2–4.	Walter of Henley.
	§§ 5–9.	Seneschaucie.
	§§ 10–19.	Walter of Henley.
	§§ 20 to end.	Unidentified.
CC. 74, 75.		Unidentified.
C. 76.	§ 1.	Seneschaucie.
	§§ 2–12.	Walter of Henley.
	§ 13.	Seneschaucie.
C. 77.		Unidentified.
C. 78.		Seneschaucie.
C. 79.	§ 1–8.	Walter of Henley.
	§§ 9 to end.	Seneschaucie.
C. 80.	§§ 1, 2.	Walter of Henley.
	§ 3.	Seneschaucie.
C. 81.	§ 1.	Walter of Henley.
	§ 2.	Seneschaucie.
C. 82.	§§ 1, 2.	Seneschaucie and Walter of Henley.
	§§ 3 to end.	Seneschaucie.
C. 83.		Unidentified.
CC. 84–88.		Seneschaucie, and much that is unidentified.

(16) Harleian, 498 B. f. 498. This copy occurs in the second of two small volumes of statutes written in a fourteenth century hand. On a fly leaf of the first volume, in a hand of the following century, is 'Per fratrem Johanem Lathbury Seniorem Liberetur fratribus minoribus de Redyng.' Walter of Henley's treatise is in a different hand from the rest of the volume, and appears to have been transcribed from a mutilated copy. It begins, 'Beau filz a leal gentz vous acoyntez,' and ends, 'Owe respondera par an xij. d. gelyne respondera par an iij. d. et a la fotz a iij. d.,' and this last sentence comes immediately after the chapter on sheep. The chapter, 'coment home despendre ses biens,' ends abruptly 'secle' (see below, p. 6), and the next begins at 'par lestente' (*ib.*). The rest of the MS. however is identical with Lansdowne, 1176, and has

the peculiarities of the Reading group; it is divided by
spaces left for initial letters, but only two headings are given,
' Coment home despendre ses biens,' and ' Pur manoirs ex-
tender vous auetz aillours mais ne mye tut.'

(20) Add. 20,709. This is a curious little book, though
it gives no important light on the text of Walter of Henley.
It is a note book dated 1571, and belonging to William
Lambarde, a celebrated antiquary of the sixteenth century,
of whom an interesting account will be found in Nichols'
Bibliotheca Topographica Britannica, I. 493. He had access
to a MS. belonging to Matthew Parker, entitled *Du Gaignage
des terres*, and made some extracts from it. This appears to
have been a MS. of Walter of Henley. One of the extracts
(f. 5) is stated to be from Walter of Henley; this gives the
estimate about ploughing as it occurs in the Canterbury
group, where the league is said to consist of twenty furlongs;
it is clear, however, that this extract was not made at the
same time as the notes from Matthew Parker's *Gaignage*,
and we are forced to surmise that two copies had been in
Lambarde's hands; one of these was the lost Parker MS.,
which probably belonged to the Luffield group, and another
was a Canterbury MS., of which a transcript and translation was made for Lambarde (see below on Oxford 21). He
also made use of the anonymous *Husbandry*; his extracts
are very meagre, but one or two remarks are worth noting.
He conjectured that the MS. he used of the *Gaignage des
terres* was written in the time of Edward I. (f. 46), and he
also explains ' that at the present day the bailiff retains his
name, but the " provost " or Præpositus is called the reeve
and the clerk is the name of the Seneschal' (f. 46).

(8) Heralds' College. Arundel MS. XIV. The volume
in which this copy of the treatise occurs is described in
the Preface to Gaimar's *Lestoire des Engles* (Rolls'
Series).

This manuscript is closely related to the *Liber Horn*, and
also to the MSS. of the Reading group; it has a table of
contents prefixed, which closely corresponds to that in the
Liber Horn, but the headings of the chapters differ from

b

those in the table, and there are some curious differences in the text. In the Introduction the second English proverb is omitted, as in the Reading group. The chapter on sheep farming ends with an unfinished sentence, which does not occur in other MSS.: ' Beau fiz si vos berbiz ou vos agneals ke vient de semeyson unt mangie de—'

There is some difference in the divisions of the text; the chapter on the Survey is divided into three; the second of the three is entitled 'Combien des acres une charue pust sustener par an,' and runs, 'E par les estendurs' (p. 6) to 'le acre de lxvi pyez de leese' on p. 8. The other divisions, beginning at ' Franc tenant' (p. 10 l. 3), 'A charue des bez' (p. 10 l. 25), and 'Al waret est une bone seson' (p. 12 l. 22), correspond to those of the Reading group rather than to those of the *Liber Horn*. The transcriber must have combined two texts, adopting the divisions of the Reading group, and giving his own headings to the chapters, but prefixing the *Liber Horn* table of contents and including the insertion on sheep farming which it contains. He, however, omitted the sentence on the amount of ploughing required on the two methods which occurs in other examples of the Reading group.

(5) GUILDHALL. I was indebted to the kindness of the Town Clerk of London for permission to examine the *Liber Horn*. The labours of the late Mr. Riley came prematurely to a close; and it is most unfortunate that he was unable to carry through the work of editing this volume. The *Walter of Henley* is of the time of Edward II., and the manuscript is of special value; the scribe has been much more careful about the headings of chapters and about the correct copying of figures than was usually the case. Though apparently some few years later than the Heralds' College MS., it represents an arrangement of the text which that transcriber had before him. In this case, however, the headings of the chapters and the divisions correspond to the table of contents. The MS. contains a long insertion on sheep farming and the sentence on the three field and two field systems.

OXFORD.

(7) Merton College CCCXXI. This is a MS. of the beginning of the fourteenth century, and is in excellent preservation. The text is similar to the *Liber Horn*; the resemblances are very close, but the differences of spelling and sometimes of construction are very curious, and suggest that this MS. was written from dictation rather than copied.

(8) Bodleian Library, Douce 98. This is a beautifully written MS. of the beginning of the fourteenth century; it is practically identical with the text printed here, as the title, divisions, and headings closely correspond with those in the Luffield MS.

(17) Digby 147. This is an early fifteenth century MS., and contains a Latin translation of Walter of Henley. The text from which this translation was made appears to be connected with that of the *Liber Horn* group, as it has similar insertions, but the chapter on the Survey is divided as in the Heralds' College MS. A new chapter, *de modo arandi*, begins at 'Ore en arrant' (p. 8 l. 12).

(21) Rawlinson MSS. B 471. This is a small volume containing various papers in the handwriting of William Lambarde. Among these is a transcript of Walter of Henley's treatise, with an interlinear translation. The transcript is not in Lambarde's writing, and the translation does not appear to be his either, but at the end are two notes written and signed by him. The first, dated May 2, 1577, is: 'Hactenus ex libro ecclesiæ Christi Cantuar. perantiquo quem scriptum fuisse autumo tempestate E. 1 vel E. 2 regis'; the second, written in December of the same year, runs: 'Tradit Joannes Balæus (in centuria Scriptorum Brytanniæ quarta fol. 304) Robertum Grosseteste, Episcopum olim Lincolniensem hunc de Agricultura Libellum de Gallico in Anglicum transtulisse. Obiit antedictus Robertus ille Anno Domini 1253, qui fuit annus regni Henrici regis tertij 37. Ut vult Matthæus parisiensis. Unde satis liquet Gualterum Henleyum, militem huiusce opusculi Authorem annis abhinc 380 aut eo plus fatis concessisse. W. Lambarde Decemb. 1577.'

Some marginal notes by Lambarde are also of interest. He has remarked the mistake peculiar to the Canterbury MSS., which give the breadth of the quarentine as 40 feet, and says: 'I think this should be 4 in breadthe, and 40 in lengthe, so that this quarentine should be all one with oure small acre.' Below this in another note, 'These names of quarentine, coture, and leuge be common in the Domesday book in the Exchequier.' The word 'messer' has been translated first as 'mower,' but this has been changed to 'overseer of husbandrie,' and Lambarde notes : 'This woord soundeth a mower, but his office was to oversee the workemen, and to kepe the cornefieldes from harme. Customes Normandie, fol. 121.'

CANTERBURY.

(10) Of the two MSS. in the Cathedral Library the later one has suffered considerably from damp. It is a fourteenth century MS., but it apparently follows the other copy, which is in a known hand, and has been identified as the writing of John de Gare, in the beginning of the reign of Edward I. This Canterbury MS. (2) is the source from which the British Museum MS. (12), as well as the Cambridge University MS. (11), and the Trinity College MS. (15) are derived.

PARIS.

This manuscript has been fully described by M. Paulin Paris in *Les Manuscrits françois de la Bibliothèque du Roi* (t. iii. p. 359), and it has also been printed by M. Lacour, *Traité inédit d'économie rurale* (Paris, 1856).

It occurs in a devotional work containing illustrations of the Scriptures, and of the legends of the saints. It was written in England in the thirteenth century, and probably for a 'grande dame.' The last eight folios of the volume contain the treatise on estate management, which consists of a curious combination of the treatise of *Walter of Henley* and the anonymous *Husbandry*. At the end several other chapters are added which contain receipts and other domestic

information. I have been unable to identify their source. The compiler of this combined treatise used a copy precisely similar to the original of the *Liber Horn*. The headings of the chapters in the *Liber Horn* are here incorporated in the text at the beginning of each chapter, and the clerk added new headings of his own. The sentence about the cost of working the two field and three field system is omitted, however.[1]

The re-arrangement has been effected without any great skill, and it is not easy to trace the reasons for the rearrangement. The chapters run as follows :—

Paris	Walter of Henley	Anonymous	Paris	Walter of Henley	Anonymous
1	p. 2		18	p. 18	
2	p. 2		19	p. 18	
3	p. 4		20	p. 22	
4	p. 10		21		p. 78
5		p. 60	22	p. 22	
6		p. 62	23		p. 78
7			24	p. 26	
8	p. 6		25		p. 76
9	p. 8		26		p. 78
10	p. 12		27		p. 72
11	p. 10		28		p. 74
12	p. 10		29	p. 28	
13		p. 60	30	p. 30	
14		p. 70	31	p. 28	
15		p. 72	32	p. 4	
16		p. 66	33	p. 32	
17	p. 16				

The last nine chapters in Lacour are unidentified.

Besides the above MSS., all of which have been examined, there are several others which came to my knowledge so recently that I have not been able to take them into account in the foregoing remarks, and I have no doubt that there are many others in private collections or elsewhere. The following may be mentioned :—

Collection of William W. E. Wynne, Esq., Peniarth, Merionethshire, No. 92. Translation of Walter of Henley's *Husbandrie*, attributed to Grosseteste. *Historical MSS. Commission, Report* II. app. 106.

---

[1] The work contains a table of English measures of land, as does the English translation, but this does not occur in any other form of the text.

Collection of the Duke of Northumberland, Syon House. D. x. 1 f. 45. *Historical MSS. Commission, Report* VI. app. 244.

Collection of Sir A. Acland-Hood, St. Audries, Somersetshire. *Historical MSS. Commission, Report* VI. app. 345.

Bodleian Library. Ashmolean MS. 1524. A fourteenth century copy, but very imperfect. Mr. Black in the catalogue of these MSS. suggests that this is more nearly the original of the Digby version than the Heralds' College MS., but it is so fragmentary that there is difficulty in speaking decidedly.

There is also a copy in Welsh, British Museum, Add. 15056, transcribed by Jolo Morganwg, *alias* Edward Williams, from a book of Thomas Hopcin, intitled 'Cato Cymraeg.'

I have also come on traces of several MSS., which appear to have been lost or destroyed in comparatively recent times.

Parker's MS., entitled *Gaynage de terres*, and used by William Lambarde, cannot be found among the Archbishop's books. A considerable portion went to Corpus Christi College, Cambridge, others to the University, and others to Lambeth, but I cannot hear of this MS. at any of these libraries.

There was also a Latin translation in a Syon College book, which was formerly in the possession of Magdalene College, Oxford, but of which I can find no traces. It commenced 'Pater aetate decrepita' according to Pegge (*Life of Grosseteste*, p. 285), and must therefore have been a different translation from that in the Digby collection, which begins 'Pater jacet in senectute.'

As a study in the process of the corruption of an author's text, the examination of these MSS. is of considerable interest. What is most remarkable is the evidence that in the thirteenth century, when the treatise was new and had no established reputation, it was found useful, and obtained a considerable circulation, but it was treated with scanty respect by the transcribers. John de Gare altered the

passage about the leagues to suit some local usages; the *Liber Horn* represents a copy which was corrupted by the insertion of the passage on sheep farming; the Luffield scribe omitted the calculations about the two field system. The copy prepared in the thirteenth century for a noble lady was recklessly mutilated, interpolated, and rearranged, and the author of *Fleta* was unscrupulous in his mode of quotation. But, fortunately, some early MS. survived untampered with, and this was more carefully dealt with by the transcribers of the Reading group in the fourteenth and fifteenth centuries.

For the present edition the Luffield MS. has been selected on three grounds : (1) Alone among the early MSS. it does not appear to have any interpolations. (2) The text is divided as it is in by far the greater number of MSS., and in the earliest MSS., for in this matter the Reading group stands alone. (8) The peculiarities of the language in which it is written give it additional interest.

4. It is unnecessary to say much of the editions of Walter of Henley. The author has been badly treated by those who have printed his work, as in no case has it been issued under his own name.

The earliest edition was the English translation issued by Wynkyn de Worde; a copy, said to be unique, is in the University Library at Cambridge, but is undated. It must have been printed from a MS. which was very closely connected with the one preserved in the Sloane collection; it has the same title, ascribing the translation to Grosseteste; it has similar divisions and headings. It is immediately followed both in the MS. and in the printed copy by a tract on the *Planting and Grafting of Trees*, adapted and translated from Palladius. At the same time there are so many minor differences of spelling and language that we cannot suppose that Wynkyn de Worde printed from the MS. still preserved.

Selden's edition of *Fleta*, containing long extracts from Walter of Henley, was published in 1647; it does not seem

necessary to add anything to what has been already said about the MS. of this work (6).

In 1856 M. Louis Lacour printed as a *Traité inédit d'économie rurale* the Paris MS. of Walter of Henley and the anonymous *Husbandry* combined. The present is the first edition which attempts to give this celebrated treatise in its original shape; the later additions have, of course, an interest of their own, and these, as well as the tracts with which Walter of Henley's was confused, are now rendered available.

IV.
THE ANONYMOUS *HUSBANDRY*.

While little can be said about the date and authorship of Walter of Henley's treatise, we have no indication at all in regard to this anonymous work, except that it had been written in time for John de Gare to transcribe it at Canterbury. It enjoyed a smaller circulation in all probability than the other, but it was treated as a practical work which it was well to possess in a handy form. The present edition has been printed, not from a book, but from a roll in S. John's College, Cambridge, which is precisely similar in shape and appearance to the rolls in which the bailiffs kept their accounts. It measures seven feet four inches long, and is about six inches wide. The hand is of the early part of the fourteenth century.

It also occurs in conjunction with Walter of Henley's treatise in the book in the Cambridge University Library marked Hh. iii. 11 f. 1653, but in a later hand than the transcript of Walter of Henley in the same book (11). It is found in the two Canterbury MSS. (2 & 10) and in the British Museum, Add. 6159, f. 217 (12); also in the Merton College MS. (7), and at Paris (4). The fact that the compiler of the Paris MS. had before him all the materials which occur in the Merton MS., and used a similar arrangement of Walter of Henley, may indicate a

close connexion between these copies. It appears that, though the number of MSS. is small, we can trace them to more than one distinct centre of transcription.

A very curious part of this treatise is the table for reducing acres based on poles of 18, 22, or 24 feet to acres based on a pole of 16 feet.[1] This must have been a difficult sum to work out, and it is to the author's credit that his arithmetic is correct, though he had evident difficulty in the addition of fractions, and expresses 1 acre $3\frac{9}{10}$ roods by the cumbrous form 'one acre and a half and a rood and a half and the sixteenth of a rood.' After all, his results give only a very rough approximation, for the recognised pole, as defined by statute, was not 16 feet but $16\frac{1}{2}$. The accurate calculation involving a square perch of $30\frac{1}{4}$ yards was probably too much for the author to attempt, and he solved the difficulty by understating the length of the statute pole; but it may be that a pole of 16 feet was used in his part of the country.

V.

SENESCHAUCIE.

There is no indication, so far as I am aware, that enables us to identify the author of the *Seneschaucie*, or to assign its date with any exactness. As already stated, it furnished the scheme for some chapters of *Fleta*, and it cannot therefore be later than the time of Edward I.; it also occurs in a MS. in Cambridge marked Mm. i. 27 f. 133, which must be assigned to the same reign. This is a quarto on parchment containing a *Registrum Brevium*, a collection of statutes, proclamations, an assise of breads of the time of King John, and many other documents. This work also occurs in the same hand as Walter of Henley in the *Liber Horn*, Dd. vii. 6 f. 50b, at Cambridge, and also in the same hand as the anonymous *Husbandry* in the fifteenth century Cambridge University MS. (Hh. iii. 11 f. 167 *bis*) which contains all these three treatises.

[1] See below, p. 68.

There are also two very imperfect copies, one at Cambridge marked Dd. ix. 38 f. 249*b*, and one in the British Museum marked Add. 5762. The curious diversity of arrangement which is found in these different copies can be best exhibited in tabular form.

Camb. Univ. Lib. Dd. vii. 6 f. 50 b	Brit. Mus. Lansdowne 559 f. 209	Camb. Univ. Lib. Hh. iii. 11 f. 167 bis	Guildhall Liber Horn	Camb. Univ. Lib. Mm. l. 27 L 133	Camb. Univ. Lib. Dd. ix. 38 f. 249 b	Brit. Mus. Add. 5762 L 131 b
seignur seneschal bailiff provost hayward charettlers charuers vachier porchier berchier daye accounters	seygnur seneschal baillif prouost hayword charetters charuers vacher porcher bercher — acunturs *end missing*	segnur seneschal bayllif prouost hayward chareters carucrs vachyr porcher bercher daye acouturs	seignur seneschal baillif provost hayward charetters carucrs vacheer porcher bercher deye acountours	seygnur seneschal baillif prouost hayward charretters caruies vacheer porcher bercher daye acunturs	— seneschalli ballivi prepositi — — — — — — — acounturcs	seignur seneschal baillif prouost — — — — — — — acountours *end missing*

VI.

GROSSETESTE'S *RULES*.

The occasion for the writing of Grosseteste's *Rules* has been so clearly discussed by Pegge that it is unnecessary to do more than state the results at which he has arrived.

'John Laci, Earl of Lincoln, a person who had a great share in the king's counsel, died July 22, 1240; and Margaret, his dowager, "had the manors of Ingoldemers, Throseby, Houton, and Seggebrock, assigned by the king for her maintenance, untill her dowry out of her late husband's lands should be set forth." She afterwards married Walter Marshall, Earl of Pembroke, 1242. And if that piece of Bishop Grosseteste's intituled " Regulæ quas bonæ memoriæ Rob. Grosseteste fecit comitissæ Lyncolniæ ad custodiendum et regendum terras, hospitium domum et familiam," was written for the service of this lady in her widowhood, it must have been composed 1240 or 1241. Robert had been but a few years then in his see, but perhaps there might be a family acquaintance and friendship in the case, John Laci having been constable of Chester,

where the bishop had been archdeacon. Margaret, again, was daughter of Robert Quinci, eldest son of Saer, Earl of Winchester, and Hawise, fourth sister of Ranulph, last Earl of Chester of that name, in whose time Grosseteste had been archdeacon there, and intimate probably in the family; and on this supposition his lordship's respect and veneration for the mother might naturally be transferred upon the daughter.' [1]

Of these *Rules* I have seen four manuscripts in French, and one in Latin. They were also translated into English, and some portions of them have been published by Mr. Brewer in the *Monumenta Franciscana*, vol. I. app. ix.[2] The translator had, however, fallen into an error which Mr. Brewer has not corrected. They were probably entitled *Reules S. Roberd*, and the translator supposed that they were rules laid down by the Bishop of Lincoln for the management of his own household and estates, whereas they appear to have been advice tendered to the Countess of Lincoln. It is of course possible that he had rules for his own household, and communicated a copy of these rules to the Countess, and that the English translation follows the bishop's rules, while the French copy here printed consists of his advice. But on the whole it seems more probable that it was a mere error on the part of the translator.

The manuscripts are as follows:

Bodleian, Douce MSS. 98 f. 128.—This copy occurs in the same volume as the treatise of Walter of Henley, and is written in the same hand; it is the one now printed.

Digby, 204 f. 3.—This is a Latin translation in a late fifteenth century hand. It begins 'Hic incipiunt regule quas bone memorie Robertus G[r]ossetete fecit comitisse Lincolniensi ad custodiendum et regendum terras osspitium et domum et familiam.' The numbers and titles of the *Rules* are not given, and the last few are wanting.

British Museum. Harl. MSS. 273 f. 81.—This volume begins with a kalendar and the Psalter in French, and most

---

[1] Pegge's *Life of Grosseteste*, 95.
[2] See also *The Babees Book* (Early English Text Soc.), p. 328.

of the contents are religious works. The copy of the *Rules* corresponds very closely with that in the Douce collection. It begins 'Ici cumencent les reules qe Robert Grosteste euesqe de Nichole fist a la cumtesce de Nichole.' The scribe has been puzzled with the calculations as to the expenditure of corn (see below, p. 127). He writes 'e ci uos despendez ij quarters le iour ce sunt xiiij la simeine ceo sunt vij cenz e x' —and here a blank space has been left, which in another hand has been filled in to make xxviij. The copy goes on 'e ci vos despendez chescun ior pur acrestre uotre aumoine ij quarters e demi ceo sunt en la symeine xvij quarters e demi en lan'—and here a blank space of a line and a half has been left; another hand has added 'ce serra ix^c quarters.' These calculations are omitted in the Latin. This MS. has been used to correct the text.

Add. 5762 f. 136.—This copy is in an early fourteenth century hand, and has a running title 'Compositio Manerii.' The headings to the rules are not given; in fact, there are no divisions of any kind. The names of Grosseteste and of the Countess of Lincoln do not appear in this copy; the treatise begins, 'Cest escrit vos aprent coment seygnur ou dame purra sauer en chescon maner totes les terres par lur parceles, totes ces rentes, custumes, seruages, vsages, franchises, fecz & ses tenementz.' The passage containing the calculations runs thus: 'Si vos despendetz deus quaters le jour ceus sunt quatortze quarters la simayne ceus sunt D.C.C. xxviij quarters par an. E si vos despendetz chescon jour pur encrestre votre aumoyne deus quarters e demy ceus sunt en la symaine xvij quarters e demy ceus sunt D.C.C. liiij [*sic*] quarters par an.'

Sloane MSS. 1986.—This fragment in English is printed as the supplement.

Canterbury.—This is an imperfect copy, in the form of a roll; but the first membrane has been lost, so that it commences in the middle of the seventeenth Rule. It is in beautiful preservation, in a thirteenth century hand, and might even have been transcribed during the author's life; the *Rules* are given, but without titles.

LE DITE DE HOSEBONDRIE

WALTER DE HENLEY

LE DITE DE HOSEBONDRIE.

Ce est le dite de hosebondrie ke vn sage homme fist iadis ke auoyt a non syre Walter de henle. E ceo fyst il pur enseyner acune gentz ke vnt teres e tenementz ke ne seuent pas toutz les poyns de hosebondrie meyntener come de gaynage de tere e de estor garder dount grans bens en poent surdre a ceus ke ceste doctrine volent entendre e apres oueryr come ci troueretz escrit.

Le pere set en sa veylesse e dyt a son fitz beu fiz uiuet sagement solom deu e solom le secle. En uers deu pensez souent de sa passion e de la mort ke ihesu crist suffry pur nus e lamet sur tote renz e ly dotet e ses comandemens tenet e gardet. Quant al secle penset de la roe de fortune coment homme mounte petit e petit en rychesse e kaunt il est al somet de la roe dunc par meschance chet petit e petit en pouerte e pus en meseyse dunt io ws pri ke solom coe ke vos teres valent par an par estente ordenet votre uie e nent plus haut. Si ws poet vos teres aprower par gaynage ou par estor ou par autre purueyance plu ke lestente le surplus metet en estu kar si ble defayle ou estor murge ou arsun sourueyne ou autre mescheances adonc ws vaudra i coe ke ws auet estue. Si ws despendet en lan la ualue de vos teres e lenpruement e vne de ces cheaunces vos surueyne vos nauez recouerir fors denprent e ke autri enprente le son gaste. Ou de fere cheuisances come acune gent ke se font marchanz[1] achatent a xx soutz e vendent a x souz. Hom dyst en reprouer ke de loyns se purueyt de pres sen joyst. Wos ueet vne gens ke vnt teres e tenemens e i ne seuent

[1] [ki.]

HUSBANDRY.

This is the treatise on husbandry that a good man once made, whose name was Sir Walter of Henley; and this he made to teach those who have lands and tenements and may not know how to keep all the points of husbandry, as the tillage of land and the keeping of cattle, from which great wealth may come to those who will hear this teaching and then do as is found written herein.

THE father having fallen into old age said to his son, Dear son, live prudently towards God and the world. With regard to God, think often of the passion and death that Jesus Christ suffered for us, and love Him above all things and fear Him and lay hold of and keep His commandments; with regard to the world, think of the wheel of fortune, how man mounts little by little to wealth, and when he is at the top of the wheel, then by mishap he falls little by little into poverty, and then into wretchedness. Wherefore, I pray you, order your life according as your lands are valued yearly by the extent, and nothing beyond that. If you can approve your lands by tillage or cattle or other means beyond the extent, put the surplus in reserve, for if corn fail, or cattle die, or fire befall you, or other mishap, then what you have saved will help you. If you spend in a year the value of your lands and the profit, and one of these chances befall you, you have no recovery except by borrowing, and he who borrows from another robs himself; or by making bargains, as some who make themselves merchants, buying at twenty shillings and selling at ten. It is said in the proverb, 'Who provides for the future enjoys himself in the present.' You see some who have lands and tenements and know not how to live.

pas viure pur quey? io le wos dirray pur coe ke eus viuent santz ordinance fere e purueyance auant mayn e despendent e gastent plus ke lur teres valent par an e kant il vnt degastes lur bens adonc ne vnt fors ke de mayn en gule e viuent en angoysse ne cheuisance ne puent fere ke bon lur seyt. Hom dit en reprouer en engleys wo þat strechet forþerre þan his wytel wyle reche in þe straue his fet he mot streche. Beau fitz sages seez en vos fes e cuntregeytet le secle ke tant est wyschous e catillous.

Atort de nului chose ne eyet ne encheson nequeret uers nully pur se bens auoyr kar hom dit en englyse Oñ ȝeer oþer to wroge wylle on honde go. Ant euere aten hende wrong wile wende. Si le gens chent en vos cours ke il seyent amercyetz par lur pers si vostre conscience vos dye ke eus seyent trop haut amerciez ke vos la amenuset issi ke vos ne seyet repris issi ne deuant deu. Alele genz vos acoyntes e amour de vos ueysins eez kar hom dit en franceys ki a bon veysyn si a bon matin. Vostre bouche sagement gardet ke vos par resun ne pusset estre repris.

[COMENT HOM DEIT DESPENDRE LES BIENS KE DEUS LUY AD DONE.][1]

Les bens ke deu vos preste sagement les gardet e despendet. En mises e en despenses deuet sauoyr quatre choses. Le vn est quant vos deuet doner coment e a ki e combyen. Le primer ke doynnet auant ke vos eez abosoyner kar meus vaudrunt ij s. deuant ke x quant hom abosoyner. Le secund si vos deuet doner ou despenses fere les facet de bone volunte e donc vos serra la chose alowe au double. E si vos donet recreaument vos perderet kant que vos imetes. Le tertz a cely donet ke vos purret valer e greuer. Le quart comben vos deuet doner ne plus ne meyns ke solum coe ke la persone est e solom ceo

---

[1] The heading to this chapter is inserted from (12).

Why? I will tell you. Because they live without rule and forethought and spend and waste more than their lands are worth yearly, and when they have wasted their goods can only live from hand to mouth and are in want, and can make no bargain that shall be for their good. The English proverb says, 'He that stretches farther than his whittle will reach, in the straw his feet he must stretch.' Dear son, be prudent in your doings and be on your guard against the world, which is so wicked and deceitful.

Have nothing from anyone wrongfully, nor seek occasion towards anyone to have his goods, for it is said in the English proverb, 'One year or two, wrong will on hand go, and ever at an end, wrong will wend.' If anyone comes into your court, let him be amerced by his peers; if your conscience tells you that they have amerced him too highly do you lessen it, so that you be not reproved here or before God. Acquaint yourself with true men and have the love of your neighbours, for it is said in the French proverb, 'Who has a good neighbour has a good morrow.' Keep your mouth prudently, that you be not justly reproved.

How a man ought to spend the wealth that God has given him.

The wealth that God lends you keep and spend prudently. In outlays and expenses you must know four things. The one is, when you ought to give, how, to whom, and how much. The first is, that you give before you are obliged to, for how much more shall two shillings be worth beforehand than ten when one is forced to give. The second is, if you must give or spend, do it with good will, and it shall be reckoned double to you, and if you give grudgingly you shall lose as much as you put out. The third is, give to him who can help and hurt you. The fourth is, how much you ought to give, neither more nor less than according to the person, and according as the business is small or

ke la bosoyne est petite ou grande ke vos auet de ly afere. Les poueres regardet ne mye pur loenge auoyr del secle mes pur loenge auoyr de deu ke vos tot troue.

[ESTENTENDEZ VOS TERRES E VOS TENEMENS PAR VOS GENS IUREY.][1]

Vos teres e vos tenemens estendet par vos leaus gentz iures. Primes estendet Cours Gardyns Columbers Cortilages ceo kil poent ualer par an outre la prise. E puy kantes acres sunt en demeyne e comben est en chescune coture e quey purrunt valer par an.[2] E quantes acres de pasture e quey valent par an. E totes autres pastures seuerales e quey valent par an. E boys ceo ke vos poez vendre sanz wast e sanz destruction e quey put valer par an outre la reprise. E francs tenans comben [3] checun tent e par queu seruice. E custumers comben checun tent e par queus seruises e custumes en deners seyent mys. E touz autres choses certeynes metez ceo ke il poent valer par an. E par les estendurs enqueret de comben hom pora semer vn acre de tere des touz maneres des bles. E combien de estor vos porret auer de sur checun manere. Par lestente purret sauoyr combyen vos teres valent par an par quey vos purret ordiner vostre uie sicum vos ay auant dit. Estre ceo si vos baylifs dient, ou vos prouos en lour acounte tant de quarters semes sur tant des acres alet ale estente e par chance vos troueret mey des acres ke eus ne vos dient. E plus des quarters semez ke mester ne fust kar vos auet a la fin de le estente de comben hom porra semer vn acre de tere de touz maneres des bles. Estre ceo si mester seyt demetre plus coustages en charues ou meyns par le estente serret sertifye. Coment? io le vos diray. Si vos teres sunt partes en treys lune partye a yuernage e lautre partye a quaremel e la terce partye a waret donc est la charue de

---

[1] Inserted from (12).   [2] [e pus kantz de pre e quey valent par an.]
[3] [de terre.]

great that you have to do with him. Have regard to the poor, not to have praise of the world, but to have praise of God, who finds you all.

SURVEY YOUR LANDS AND TENEMENTS BY YOUR SWORN MEN.

Survey your lands and tenements by true and sworn men. First survey your courts, gardens, dove-houses, curtilages, what they are worth yearly beyond the valuation; and then how many acres are in the demesne, and how much is in each cultura, and what they should be worth yearly; and how many acres of pasture, and what they are worth yearly; and all other several pastures, and what they are worth yearly; and wood, what you can sell without loss and destruction, and what it is worth yearly beyond the return; and free tenants, how much each holds and by what service; and customary tenants, how much each holds and by what services, and let customs be put in money. And of all other definite things put what they are worth yearly. And by the surveyors inquire with how much of each sort of corn you can sow an acre of land, and how much cattle you can have on each manor. By the extent you should be able to know how much your lands are worth yearly, by which you can order your living, as I have said before. Further, if your bailiffs or provosts say in their account that so many quarters have been sown on so many acres, go to the extent, and perhaps you shall find fewer acres than they have told you and more quarters sown than was necessary. For you have at the end of the extent the quantity of each kind of corn with which one shall sow an acre of land. Further, if it is necessary to put out more money or less for ploughs, you shall be confirmed by the extent. How? I will tell you. If your lands are divided in three, one part for winter seed, the other part for spring seed, and the third part fallow, then

tere ix^{xx} acres. E si vos teres sunt party en deus com sunt en plusurs pays lune meyte seme a yuernage e a quaremel e lautre meyte a waret adonc serra la charue de tere viij^{xx} acres. Alet ale estente e veet comben des acres vos auet en demeyne e la purret estre sertefie.[1]

Acune gent dirrunt ke la charue ne put pas sustenir par an viij^{xx} acres ne ix^{xx} acres.[2] E io vos mustray ke si put. [Byen sauet ke vne coture deyt estre de quarante perches de long, e iiij perches de lee. E la perche le rey est de xvi pez e demi e adonc est le acre de lxvj pyez de lcyse. Ore en arrant alet xxx[3] foys entur pur fere le reon plus estreyt e kant le acre ert pararre a donkes estes ale lxxij coutures ke sunt vj lywes kar ceo fet asauoyr ke xij cotures sunt vne lywe.][4] Mout sereyt poure le cheval[5] ou le boef ke ne put aler du matyn belement le pas iij[6] lywe de voye de son rescet e retorner a nonne. E par autre reyson vos moustruy io ke ele put ben tant fere. Byen sauet ke il y a en lan lij semaynes ore ostet viij semaynes pur feyretz e pur autres desturbances[7] a donc demorent xliiij semaynes ouerables. E en tot cel tens nauera la charue a fere a larrue de waret e alarrure de semaysl de yuernage e de quaremel fors ke a la iornee iij rodes e demy rode e au rebyner un acre.[8] Ore veet si vne charue ke fust adreyt garde e sywy si ele ne pust tant fere a la iornee. E si uus auet tere ou estor put estre metet peyne de estorer le solum ceo ke la tere de-

---

[1] [Autre taunt auera a fere la charuwe de viij^{xx}. acres com la charuwe de ix^{xx} acres. Le uolez vos veer. Quant a viij^{xx} acres prenez xl acres a yuernage e xl acres a quaremel e iiij acres a waret returnez e rebinez les iiij^{xx} e dunkes ira la charue xjj^{xx} acres. En dreit de la charuwe de ix^{xx} acres lx acres a yuernage e a quaremel e lx a waret e puys returnez e rebinez les lx acres a dunk ira la charuwe en lan xij^{xx} acres.]

[2] [e ieo vous di pur veirs ke si la charuwe seyt garde e seruie a son dreyt kele pura tant fere.]

[3] xxxvi.

[4] [Bien sauez ke vne quarenteyne deit auer quaraunte perches en longur e xl en leyse e la perche le roy est de xvi pez e demy e donk ad le acre en leyse xlvj peez. E quaunt vos auez fet vostre torn xxxij fez en un reon de vn pe de lee a donke est lacre arre mes pur auer le reon plus estreit fetes vostre torn quaraunte fiez en tors pus estes ale iiij^{xx} coteres ki amountent iiij lywes kar quarante perches en longur funt une coture e vint cotures funt une lywe.]

[5] affre.   [6] dou.

[7] [ke pussent suruent.]

[8] [le iour.]

is a ploughland nine score acres. And if your lands are divided in two, as in many places, the one half sown with winter seed and spring seed, the other half fallow, then shall a ploughland be eight score acres. Go to the extent and see how many acres you have in the demesne, and there you should be confirmed.

Some men will tell you that a plough cannot work eight score or nine score acres yearly, but I will show you that it can. You know well that a furlong ought to be forty perches long and four wide, and the king's perch is sixteen feet and a half; then an acre is sixty-six feet in width. Now in ploughing go thirty-six times round to make the ridge narrower, and when the acre is ploughed then you have made seventy-two furlongs, which are six leagues, for be it known that twelve furlongs are a league. And the horse or ox must be very poor that cannot from the morning go easily in pace three leagues in length from his starting-place and return by three o'clock. And I will show you by another reason that it can do as much. You know that there are in the year fifty-two weeks. Now take away eight weeks for holy days and other hindrances, then are there forty-four working weeks left. And in all that time the plough shall only have to plough for fallow or for spring or winter sowing three roods and a half daily, and for second fallowing an acre. Now see if a plough were properly kept and followed, if it could not do as much daily. And if you have land on which you can have cattle, take pains to stock it as the land requires. And know for truth if you

mande. E sachet pur voyrs si vos estes ben estore e votre estor seyt garde e gwye, il respondra al terz de la tere par estente. Franc tenant ou custumers si eus dedient seruices ou custumes a lestente verret la certeynte.

DE SAUOYR ESLYRE VOS SERIANZ.

Si baylyf ou seriant deuet eslyre ne les elyset pur parente ne pur especiaute ne autres si il ne seyent de bon renon e ke il seyent leauz e auertitz e ke eus sachent de estor e de gaynage. Ne prouez ne messers ne neet fors de vos hommes demeyne si vos les eet e ceo par election de vos homages kar si eus trespassent de eus aueret recoueryr.

DE SURUER VOS OUERAYNNES.

Al commencement de wareter e de rebyner e de semer ke le baylyf e le messer ou le prouost seyent enterement oue les charues a ueer ke eus facent ben e plenerement lur ouerayne. E a la fin de la iornee veent comben eus ount fet e de taunt respoynent checun ior apres si eus ne sachent demoustrer certeyne deturbance. E pur ceo ke serianz de custume relinquissent en lur ouerayne il est mester de contrewayter lur fraude. Estre ceo il est mester ke hom [1] les sourueye souent.[2] E de autre part le baylyf deyt sourueyer touz ke eus ben facent. E ci eus ne facent ben ke eus seyent repris.[3]

A charue des bez od deus cheuaus vos treet plus tot ke a charue tut de cheuaus si la tere ne seyt si perouse ke buefs ne se pussent eyder des pes pur quey? Io le vos diray le cheual custe plus ke le buef. Estre ceo la charue des buefs irra atant en lan come la charue des cheuaus pur ceo ke la malyce des charuers ne suffrent mye la charue [4] aler hors del pas nent plus ke la charue des buefs.

[1] le messer. [2] chescun or. [3] [e chastiez.] [4] [des cheuaus.]

are duly stocked, and your cattle well guarded and managed, it shall yield three times the land by the extent. If free tenants or customary tenants deny services or customs you will see the definite amount in the extent.

To know how to select servants.

If you must choose a bailiff or servant, do not choose them for kindred or liking, or other reasons, if they are not of good reputation, and let them be true and prudent and know about cattle and tillage. Have no provosts or messers except from your own men, if you have them, and that by election of your tenants, for if they do wrong you shall have recovery from them.

Of overseeing your labourers.

At the beginning of fallowing and second fallowing and of sowing let the bailiff, and the messer, or the provost, be all the time with the ploughmen, to see that they do their work well and thoroughly, and at the end of the day see how much they have done, and for so much shall they answer each day after unless they can show a sure hindrance. And because customary servants neglect their work it is necessary to guard against their fraud; further, it is necessary that they are overseen often; and besides the bailiff must oversee all, that they all work well, and if they do not well let them be reproved.

With a team of oxen with two horses you draw quicker than with a team all horses, if the ground is not so stony that oxen cannot help themselves with their feet. Why? I will tell you: the horse costs more than the ox. Besides a plough of oxen will go as far in the year as a plough of horses, because the malice of ploughmen will not allow the plough [of horses] to go beyond their pace, no more than the plough of

Estre ceo en graunt duresce la ou la charue des cheuaus serra arestu la charue des befs passera. E le volet vos veer comben le cheual couste plus ke le buef? Io le vos dyray. Custume est edreyt ke bestes des charues seyent a la croche entre la foste de seynt luc e la feste de la seyt croys en may par vint e cynk semeynes. E si le cheual deyt estre enpoynt pur fere sa iornee il ly couent auer al meyns la nut le sime part de vn bussel de aueyne le pris de mayle e al meyns duze derees de herbage en este. E chescune semeyne ke vn plus vn autre meyns vn dener en ferrue si yl deyt estre ferre des quatre pecz la summe est xij s. v d.[1] on lan saunz forage e payle.

Coment vos deuet buefs gardyr.

E si le buef deyt estre en poynt a fere son ouerayne adonc couent ke il eyt al meyns tres garbez e demi de aueyne la semayne le pris de vn dener. E x garbes de aueyne respondrunt de vn bussel de aueyne par eyme e en la seson de este xij deree de herbage la summe est iij s. i d. saunz forage o payle. E quant le cheual est viel e recreu donc ni ail for le quir. E quant le buef est viel od x derrees de herbage vaudra au larder ou auendre tant com il coustera. Al waret est vne bone seson en aueryl si la terre depiece apres la charue. E al rebiner apres la seyn Iohan kant la poudre leue apres la charue. E al arrer al semayl quant la terre est assise e nest pas trop croyz mes ki a mut afere ne put pas saueyr[2] totes les bones seysons. E quant vos waretter si vos trouet par fount bone terre adonc arret vn reon quarre pur auoyr de la bone terre repose mes ne atamet mye la mauueyse tere. E arret nettement issi ke ne demurge couvert ne descouert. Ne rebynet mie trop parfount mes ke vos pusset destrure les cardons[3] escarsement kar si la terre seyt rebyne parfount e la terre seyt enbu de ewe qant hom deyt aler[4] al semayl la charue ne porra ateyndre a nule certeyne tere mes va flotant com en bowe.

[1] [e ob.]   [2] atendre.   [3] [e le herbage.]   [4] arrer.

oxen. Further, in very hard ground, where the plough of horses will stop, the plough of oxen will pass. And will you see how the horse costs more than the ox? I will tell you. It is usual and right that plough beasts should be in the stall between the feast of St. Luke and the feast of the Holy Cross in May, five-and-twenty weeks, and if the horse is to be in a condition to do his daily work, it is necessary that he should have every night at the least the sixth part of a bushel of oats, price one halfpenny, and at the least twelve pennyworth of grass in summer. And each week more or less a penny in shoeing, if he must be shod on all four feet. The sum is twelve shillings and fivepence in the year, without fodder and chaff.

How you must keep your oxen.

And if the ox is to be in a condition to do his work, then it is necessary that he should have at least three sheaves and a half of oats in the week, price one penny, and ten sheaves of oats should yield a bushel of oats in measure; and in summer twelve pennyworth of grass: the sum three shillings, one penny, without fodder and chaff. And when the horse is old and worn out then there is nothing but the skin; and when the ox is old with ten pennyworth of grass he shall be fit for the larder, or will sell for as much as he cost. April is a good time for fallowing, if the earth breaks up after the plough; and for second fallowing after St. John's Day, when the dust rises behind the plough; and for ploughing for seed when the earth is firm and not too cracked. But he who has much to do cannot wait for all the good seasons. And when you fallow, if you find good earth deep down, then plough a square ridge, to let the good land rest, but do not cut off the bad land; and plough cleanly, so that none remains covered or uncovered. At second fallowing do not go too deep, but so that you can just destroy the thistles, for if the earth is ploughed too deep at second fallowing, and the earth is full of water, then when one must plough for sowing the plough shall reach no sure ground, but

E si la charue puet aler ii deyes plus parfount ke la tere ne fust rebyne adonc trouereyt la charue certeyne tere e se nettireyt e delyuerreyt de bowe e freyt belle arrure e bon.

De tenyr le reon.

Al semer ne alet¹ mye large reon fors ke petit e ben ioynt ensemble ke la semence pusse cheyr owel. E si vos arret large reon pur ben espleyter vos fret damage. Coment? io vos dirray quant la tere ert seme donc vendra la herce e sakera le ble iekes en les croes ke est entre les deus reons e le reon ke est large serra descouert ke ren ny crester du ble. E le volet vos veer? quant le ble est de suz terre alet au chef² e vos verret ke io vos dy voyre. E la terre ke deyt estre seme de sus reon ueet ke il seyt arre de menu reon c la terre en hauce tant com vos purret. E ueet ke le reon ke est entre les deus seylloyns seyt estreyt. E la terc ke gyst come ceo fust vne crest en cel reon de souz le pye senestre apres la charue ke seyt reuerse e adonc serra le reon asset estreyt.

De semer vos teres.

Vos terres semet par tens issi ke la terre seyt assys e les bles en racines auant le fort yuer. Si auenture auéyne ke un grant pluye veyne ou chete sur la tere de denz les viij iours ke ele seyt seme e pus veyne vn aspre gele e se teyne deus iors ou treys si la tere seyt croez le gele percera la tere tant parfount come le ewe entre e par tant le ble ke est germy e tendre mut serra pery. Deus maneres teres ke sunt al quaremel semet par tens tere arzilouse e tere perouse pur quey? io le vos dirray. Si la seson seyt sek en le tens de marz dunc

---

¹ arrer.   ² [de la cotere e regardez le ble uers lautre chef.]

goes floundering, as in mud. And if the plough can go two finger-lengths deeper than at second fallowing, then the plough will find sure ground, and clear and free it from mud, and make fine and good ploughing.

To keep the ridge.

At sowing do not plough large furrows, but little and well laid together, that the seed may fall evenly; if you plough a large furrow to be quick you will do harm. How? I will tell you. When the ground is sown, then the harrow will come and pull the corn into the hollow which is between the two ridges, and the large ridge shall be uncovered, that no corn can grow there. And will you see this? When the corn is above ground go to the end of the ridge, and you will see that I tell you truly. And if the land must be sown below the ridge see that it is ploughed with small furrows, and the earth raised as much as you are able. And see that the ridge which is between the two furrows is narrow. And let the earth which lies like a crest in the furrow under the left foot after the plough be overturned, and then shall the furrow be narrow enough.

To sow your lands.

Sow your lands in time, so that the ground may be settled and the corn rooted before great cold. If by chance it happens that a heavy rain comes or falls on the earth within eight days of the sowing, and then a sharp frost should come and last two or three days, if the earth is full of holes the frost will penetrate through the earth as deep as the water entered, and so the corn, which has sprouted and is very tender, will perish. There are two kinds of land for spring seed which you must sow early, clay land and stony land. Why? I will tell you. If the weather in March should be dry, then the ground will harden too much and

en durcist la terc¹ trop e la terc perouse se ensechit mes ele
ouert par quey il est mester ke ceste tercs seyent par tens
semes issi ke les bles pussent estre noriz par la seue de iuer.

Pur delyuerer teres de cretyne.

Les teres creyous e sabclous nest pas mester de semer
si par tens kar ceo sunt dcus dereyus ke sunt eschiouz
dc estre reuerse en grant moysture mes al semayl ke la
tere seyt vn poy arose. E quant vos teres sunt semes les
teres de mareys e les teres ewouses les fetes ben reoner
e le cours del ewe fetes coure² issi ke les teres de ewe se
pussent delyueryr. Vos blez fetes munder e sarcher apres
la seynt Ioan kar deuant nest pas seson bone. Si vos
trenchet les cardons quinze iors ou viij iours dcuant la
seynt Ioan de checun vendrunt deus ou treys. Vos blez
facet sagement taylyr e en grange entrer.

A fere le issue de la grange.

Al issue de la grange metet vn leal home en vos aflet ki
put charger le prouost lcaument kar hom veyt souent ke
le granger e le gerneter se ioyne en semble pur mal fere.
Vos prouos e vos gerneters fetes karker³ mesures issi kc
al vtime bussel demurge vn cauntel pur le gast ke chet al
entrer e al yssir du gerner kar al coumble est fraude.
Coment ? io le vos diray quant lc prouost a rendu aconte
de issue de grange donc fetes prouer le bussel dunt il fu
carke. Si le bussel est large vos troueret ke quatre coum-
bles frunt le quint ou poy plus oy poy meyns. E si il
est meyns large de cynk le syme. E si il est meyns large
de sis le setime. E si il est vncore meyns large de set le
vtime. E uncore si il est meyns large de vyt le nefimc.
E issi de checun ou poy plus ou poy meyns. Ore venent
aucuns de ces prouos cn ne rendent aconte fors de uit le
syme quel ke le bussel seyt large ou estreyt. E si le bussel

---

¹ [arsillose.]   ² enlarger.   ³ [par dreytes.]

the stony ground become more dry and open, so it is necessary that such ground be sown early, that the corn may be nourished by the winter moisture.

To free lands from too much water.

Chalky ground and sandy ground need not be sown so early, for these are two evils escaped to be overturned in great moisture, but at sowing let the ground be a little sprinkled. And when your lands are sown let the marshy ground and damp ground be well ridged, and the water made to run, so that the ground may be freed from water. Let your land be cleaned and weeded after St. John's Day; before that is not a good time. If you cut thistles fifteen days or eight before St. John's Day, for each one will come two or three. Let your corn be carefully cut and led into the grange.

To make the issue of the grange.

When the stock of the grange is taken, place there a true man in whom you trust, who can direct the provost rightly, for one often sees that the grange-keeper and barn-keeper join together to do mischief. Make your provost and barn-keepers fill the measures, so that for every eight bushels a cantle shall be left for the waste which takes place at the putting in and taking from the barn, for in the comble is fraud. How? I will tell you. When the provost has rendered account for the return of the grange, then cause the bushel which he filled with to be proved. If the bushel be large then four heaped up will make five, more or less; if it be smaller five will make six; if smaller six will make seven; if still smaller eight will make nine, and so on for each, more or less. Now some of these provosts will only render account for eight in the seam, whether the bushel be large or small,

seyt large il ia deseyte grant. Si lissue de votre grange ne respoyne fors al ters de votre semayl vos ny gaynet ren si ble ne se vende byen.

DE COMBEN VOS SEMERET VN ACRE DE TERE.

Byen sauet ke vn acre seme aforment prent treys arrures hors pris teres ke sunt semes checun an. E quey vn plus vn autre meyns checune arrure vaut vi d. E le hercer vaut i d. E sur lacre couent il semer al meyns deus bussels ore valent a la seynt mychel al meyns les ij bussels xij d. E le sercler [1] vne mayle. E le syer v d. o le carier en aust i d. e le forage aquiterat le batre. Al tierz de votre semayl donc deuet avoyr vi bussels e valent iij s. e vos custages amontent iij s. e iij ob. e la terre est uotre e ne mye alowe.

COMENT VOS DEUET VOSTER SEMENCE CHANGER.

Changet voster semence checun an a la seynt michel kar plus vos aprowera la semence ke est cru sur autre terre ke ne fust cele ke est cru sur la terre meymes. E le volet vos veer ? fetes arrer deus seyluns en iur e semet le vn de la semence ke est achate e lautre semet de ble ke vos est creu al aust e vos verret ke io vos dye veyr.

COMENT HOM DEYT FENS GARDER E NORIR.

Vostre estuble ne vendet ne de la tere ne le remuet si vos neet mester de couerir mesuns si vos le remuet par le meyns vos perderet le plus.[2] Beau fiz feens facet enhaucer e oue la terre medler. E votre berchere facet checune quinseyne marler de terre arzylouse ou de bone tere cum

---

[1] [vaut.]   [2] See Introduction, p. xxv.

and if the bushel be large there is great deceit. If the return of your grange only yields three times the seed sown you will gain nothing unless corn sells well.

For how much you shall sow an acre.

You know surely that an acre sown with wheat takes three ploughings, except lands which are sown yearly; and that, one with the other, each ploughing is worth sixpence, and harrowing a penny, and on the acre it is necessary to sow at least two bushels. Now two bushels at Michaelmas are worth at least twelvepence, and weeding a halfpenny, and reaping fivepence, and carrying in August a penny; the straw will pay for the threshing. At three times your sowing you ought to have six bushels, worth three shillings, and the cost amounts to three shillings and three halfpence, and the ground is yours and not reckoned.

How you ought to change your seed.

Change your seed every year at Michaelmas, for seed grown on other ground will bring more profit than that which is grown on your own. Will you see this? Plough two selions at the same time, and sow the one with seed which is bought and the other with corn which you have grown: in August you will see that I speak truly.

How you ought to keep and prepare manure.

Do not sell your stubble or take it from the ground if you do not want it for thatching; if you take away the least you will lose much. Good son, cause manure to be gathered in heaps and mixed with earth, and cause your sheepfold to be marled every fortnight with clay land or with good earth, as the cleansing out of ditches, and then strew it over.

descurement de forsses e pus estramer sure. E si forge demurge outre la prise de sustenance de votre estor le facet estramer dedens la curt e de hors en wascels. E votre berchere facet encement estramer e vos faudes ensement. E auant ke la sekeresse de marz veyne vos fenz facet venyr ensemble ke sunt esparpylez en la curt e de hors. E quant vos deuet marler ou fens caryer eet vne home de ky vos fyet ke seyt outre les carecters le primer ior e ke il veye ke eus facent ben lur ouerayne e sanz feyntyse. E ala fyn de la iornee veyet cumben eus vnt fet e de tant respoynent checune iornee si eus ne sachent mustrer certeyne desturbance. Vos fenz ke sunt medles od tere les metet sur tere sabelouse si vos le hauet pur quey? io le vos dirray. Le tens deste est chaut e le sabelun est chaut e les fens sunt chaus. E kant les treys chalynes sasemblent par la grant chalyne si flestrysent apres la seynt Ioan les orges ke cressent en sabelun cum vos purret veer la[1] vos alet par pays en plusurs lyus. Au uespre la tere ke est medle od fens refreydyst le sabelon e noryt vne rosee e partant est le ble meuz sauue. Vos teres femet e ne les arret mye trop parfount pur ceo ke fens gastent en descendant. Ore vos dyray quel auantage vos aueret des fens ke sunt medles oue tere si les fens fussent pur de eumeymes il dureynt deus anz ou treys solum ceo ke la tere est freyde ou chaude. Les fens medles oue tere vos durunt al duble mes il ne serrunt mye si poynanz. Byen sauet ke marle dure plus ke fens pur quey? pur ceo ke fens wastent en descendant e marle en amuntant. E pur quey durent les fens medles plus longement ke les fens purs? io le vos dyray. Les fens e la tere ke sunt aers ensemble la tere sustent las fens ke il ne poent gaster endenscendant tant cum dusent naturelment pur quey io vos diray ke fenz facet noryr solum votre pouer. E vos fens quant sunt esparpylez e vn poy arosez adonc est seson ke il seyent reuerses adonc la tere e les fens se prouent de meuz ensemble. E si vos metet vos fens sur waret il serrunt tut le plus au rebyner reuerse

[1] [ou.]

And if fodder be left beyond that estimated to keep your cattle cause it to be strewed within the court and without in wet places. And your sheep-house and folds also cause to be strewed. And before the drought of March comes let your manure, which has been scattered within the court and without, be gathered together. And when you must cart marl or manure have a man in whom you trust to be over the carters the first day, that he may see that they do their work well and without cheating, and at the end of the day's work see how much they have done, and for so much must they answer daily unless they are able to show a definite hindrance. Put your manure which has been mixed with earth on sandy ground if you have it. Why? I will tell you. The weather in summer is hot, and the sand hot and the manure hot; and when these three heats are united after St. John's Day the barley that grows in the sand is withered, as you can see in several places as you go through the country. In the evening the earth mixed with manure cools the sand and keeps the dew, and thereby is the corn much spared. Manure your lands, and do not plough them too deeply, because manure wastes in descending. Now I will tell you what advantage you will have from manure mixed with earth. If the manure was quite by itself it would last two or three years, according as the ground is cold or hot; manure mixed with earth will last twice as long, but it will not be so sharp. Know for certain that marl lasts longer than manure. Why? Because manure wastes in descending and marl in ascending. And why will manure mixed last longer than pure manure? I will tell you. Of manure and the earth which are harrowed together the earth shall keep the manure, so that it cannot waste by descending as much as it would naturally. I tell you why, that you may gather manure according to your power. And when your manure has been spread and watered a little, then it is time that it should be turned over; then the earth and the manure will profit much together. And if you spread your manure at fallowing it shall be all the more turned over at second ploughing, and

desouz tere e au semayl regete¹ amount e od la tere medle. E si il mys sur le rebyner au semayl est tot le plus de souz tere² epoy medle oue la tere e ceo nert pas prou. E la [faude]³ tant com plus pres est du semayl meuz vaudera. E a la feste notre dame la premere fetes enoyter vostre faude solum ceo ke vos aueret berbys ou plus ou meyns. Car en cele ceson gettent mut de fens.

Coment vos deuet voster estor trier.

Voster estor vne feyz en lan fetes trier entre pasche e pentechoste ceo est asauer buefs⁴ vaches⁵ aumayl e tels ke ne sunt mye a retenyr, ke yl seyent mys pur engressyr. E si vos mettet custage pur engressyr en herbage vos igayneret. E sachet pur voyr plus couste le malueys ke le bon. Coment? io le vos dyray. · Si ceo beste ouerable il couent estre regarde plus ke autre e plus esparnye e de ceo ke il est esparnie les autres sunt greues par sa defaute. E si vos deuet estor achater, le achatet entre pasche e penthechoste car adonc sunt bestes megres e bon marche. E vos cheuaus changet auant ke seyent trop veuz e recruz ou mahaynes car de poy de custage purret releuer bons e iuuenes si vos vendet achatet en tens e en seson. Coment hom deyt estor garder nest pas mal si vos le sauet pur les vos en senser. E quant eus verunt ke vos ensauet eus se penerunt le plus a ben fere.

Coment vos deuet vos bestes de la charue garder.

Vos bestes de la charue deuet garder keus eyent pasture suffisuntment affere lur ouerayne e ke eus ne seyent trop mys a desuz quant eus vendrunt a la charue.⁶ Car vos metteret trop de custage del releuer. Estre ceo votre gaygnage serra arey. En mesuns ne les metet mye en

¹ reuertiz.
² [reuersis.]
³ [terre faudeye.]
⁴ [de la charuwe.]
⁵ [e vos chiuals charetters.]
⁶ cretche.

at sowing shall come up again and be mixed with earth. And if it is spread at second ploughing at sowing it is all the more under the earth and little mixed with it, and that is not profitable. And the nearer the fold is to the sowing the more shall it be worth. At the first feast of our Lady enlarge your fold according as you have sheep, either more or less, for in that time there is much manure.

How you ought to inspect your cattle.

Sort out your cattle once a year between Easter and Whitsuntide—that is to say, oxen, cows, and herds—and let those that are not to be kept be put to fatten; if you lay out money to fatten them with grass you will gain. And know for truth that bad beasts cost more than good. Why? I will tell you. If it be a draught beast he must be more thought of than the other and more spared, and because he is spared the others are burdened for his lack. And if you must buy cattle buy them between Easter and Whitsuntide, for then beasts are spare and cheap. And change your horses before they are too old and worn out or maimed, for with little money you can rear good and young ones, if you sell and buy in season. It is well to know how one ought to keep cattle, to teach your people, for when they see that you understand it they will take the more pains to do well.

How you ought to keep your beasts for the plough.

You must keep your plough beasts so that they have enough food to do their work, and that they be not too much overwrought when they come from the plough, for you shall be put to too great an expense to replace them; besides, your tillage shall be behindhand. Do not put them in

pluous tens. Car vn eschaufure veynt entre le quyr e la pel e entre le quir e la leyne ke turne a grant damage as bestes. E si vos estor eyent prouendre decustume ke lur seyt done de cler iur a vn de messers ou del prouost e medle o vn poy de [orge¹] pur ceo ke il ia trop arrestes e houireynt les bouches as cheuaus. E pur quey lur doryet vos par testmoynye e od payle? io le vos diray pur ceo ke il auent souent ke les bouers emblent la prouendre e les cheuaus manguent de meuz le payle pur la prouendre e engrossiscnt e beyueynt le meuz. E forage as buefs ne lur seyt pas done grant quantite a vne foys mes poy e souent a donc maguent ben e poy gastent. E quant il ia grant quantite deuant eus il manguent lur saule e pus seent e roungent e par la sufflure de lur aleyne le forage comencent en hayr e le gastent. E les estor ke il seyent waez e quant sunt ensechys conreyez car ceo lur fet grant ben. E les bufs ke il seyent conreyes de vn torcaz le iur e par tant se leschirunt demeuz. E vos vaches ke eus eyent suffisante pasture ke le blanc ne seyt arrery.² E quant le vel madle est veele ke il eyt son let enterement vn moys al chef del moys ly tollet vn treon e de semayne en semayne vn treon adonc letera viij semaynes. E metet a manger deuant luy ke il pust aprendre de manger. E la femele eyt son let enterement treys semaynes e luy tolet les autres treons si com le madle. E ke il eyent del ewe en tens de sekison³ de dens mesons e de hors car plusurs murent en la tere de la maladye del pomun par defaute de ewe. Estre ceo si il ia nul beste ke comence a descheyer metet custage asusteyner le.⁴ Car hom dyt en reprouer Ben eyt le dener ke sauue deus.

¹ [payle de forment ou de aueyne e ne mie od paile de orge.]
² amenuse.
³ seuerison.
⁴ [ke il ne murge pur defaute de aye.]

houses in wet weather, for inflammation arises between the skin and the hair and between the skin and the wool, which will turn to the harm of the beasts. And if your cattle are accustomed to have food, let it be given at midday by one of the messers or the provost, and mixed with little barley, because it is too bearded and hurts the horses' mouths. And why shall you give it them before some one and with chaff? I will tell you. Because it often happens that the oxherds steal the provender, and horses will eat more chaff for food and grow fat and drink more. And do not let the fodder for oxen be given them in a great quantity at a time, but little and often, and then they will eat and waste little. And when there is a great quantity before them they eat their fill and then lie down and ruminate, and by the blowing of their breath they begin to dislike the fodder and it is wasted. And let the cattle be bathed, and when they are dry curry them, for that will do them much good. And let the oxen be curried with a wisp of straw every day, and thereby they will lick themselves more. And let your cows have enough food, that the milk may not be lessened. And when the male calf is calved let it have all the milk for a month; at the end of the month take away a teat, and from week to week a teat, and then it will have sucked eight weeks, and put food before it, that it may learn to eat. And the female calf shall have all the milk for three weeks, and take from it the teats as with the male. And let them have water in dry weather within the houses and without, for many die on the ground of a disease of the lungs for lack of water. Further, if there be any beast which begins to fall ill, lay out money to better it, for it is said in the proverb, 'Blessed is the penny that saves two.'

COMBEN VOS VACHES DEYUENT RESPONDRE DE BLANC.

Si vos vaches seyent tries issi ke les malueyses seynt ostes e vos vasches seynt puez en pasture de mareys salync donc deyuent deus vaches respondre de vne peyse de furmage¹ e de demy² galon de bure la semayne. E si il seynt peuz en pasture de boys ou en pres apres fauchisons ou en estuble donc deyuent treys vaches respondre de vne peyse de furmage e de demy galon de bure la semayne entre pasche e la seynt michel sanz rewayn. E xx mere berbyz ke sunt peuz en pasture de mareys salyne dey e ben poent respondre de furmage e de bure si cum les ij vaches auant nomes. E si vos berbyz seynt peuz de freche e de waret donc deyuent xxx mere berbyz respondre de bure e de formage sicum les treys vaches auant nomes.³ Ore en ia il plusurs serianz e prouoz e dayes ke contredyrunt ceste chose e ceo est par la reson ke eus dounent e gastent e manguent del blanc. E sachet pur veyr le blanc ne seyt despendu ne gaste aylurs ke en la chose meymes de tant deyuent e ben poent respondre. Car io lay esproue. E le volet vos veer endreyt des treys vaches ke deyuent fere vne peyse pouere serreyt vne de celes treys vaches de ky hom ne put auer en deus iurs vn formage ke vausist mayle ceo cerreyt en vi iurs treys formages le pris de treys mayles. E le setime iur eydera a la disme e al wast ke va par en coste. Ore qui serreynt ceo treys mayles en xxiiij semaynes ke sunt entre pasche e la seynt mychel ceo sereyt treys souz. Ore metet la secunde vache autant. E la terce autant a donc aueret ix souz e pur tant purret vos auoyr vne peyse de formage a comune uente. Ore serreyt poueres vne de celes treys vaches de ky hom neust le terz de vn potel de bure la semayne. E si le galun de bure vaut vj d. donc vaut le terz de vn potel i d.

---

¹ [de la pasche dekes la saint michel.] ² un.
³ See Introduction, p. xxv.

How much milk your cows should yield.

If your cows were sorted out, so that the bad were taken away, and your cows fed in pasture of salt marsh, then ought two cows to yield a wey of cheese and half a gallon of butter a week. And if they were fed in pasture of wood, or in meadows after mowing, or in stubble, then three cows ought to yield a wey of cheese and half a gallon of butter a week between Easter and Michaelmas without rewayn. And twenty ewes which are fed in pasture of salt marsh ought to and can yield cheese and butter as the two cows before named. And if your sheep were fed with fresh pasture or fallow, then ought thirty ewes to yield butter and cheese as the three cows before named. Now there are many servants and provosts and dairymaids who will contradict this thing, and that is because they give away and waste and consume the milk ; and know for certainty the milk is not wasted otherwise but in the same thing, for so much they ought to and can yield, for I have proved it. And you will see it with regard to the three cows that ought to make a wey. One of these cows would be poor, from which one could not have in two days a cheese worth a halfpenny ; that would be in six days three cheeses, price three halfpence. And the seventh day shall help the tithe and the waste there may be. Now that will be three halfpence in twenty-four weeks which are between Easter and Michaelmas—that is, three shillings. Now put as much for the second cow, and as much for the third, and then you will have nine shillings, and thereby you have a wey of cheese by ordinary sale. Now one of these three cows would be poor, from which one could not have the third of a pottle of butter a week, and if the gallon of butter is worth sixpence then is the third of a pottle worth a penny.

Coment vos deuet vos pors trier.

Vos pors fetes trier vne fyez en le an e si vos trouet nul ke ne seyt pas seyn le remuet. Veres des[1] truyes neet si eus ne seyent de bon lyn vos autres pors femeles fetes sauer ke eus perdent le porceler donc vaudra le bacon autant com del madle. E ke eus pussent auer power de foyuer. En treys meys aueront il mester de eyde. En feuerer. En marz. En aueryl. En treys feys par an deyuent trues porceler si ceo ne seyt par male garde. Vne noreture est as pors de auer longe matinee e de gysyr sek. Vos purcels fetes ensauer tant com eus letent adonc cresterunt le meus.

Coment vos deuet vos berbyz trier.

Veet ke voster bercher ne seyt pas yrrous kar par vn irre acun peust estre vilement chace dount ele put estre pery. La ou les berbetz vount pessant e le bercher voyst entre a eus.[2] Vos berbytz fetes trier vne foys par an entre pasche e pentechoste e ceus ke ne sunt mye a retenyr les fetes par tens tondre e mercher des autres e les metet en boys ke seyt enclos ou en autre pasture ou il pussent engressyr o entur la seynt Iohan les vendet car donc serra char de motun en seson. E la leyne de ceus seyt vendu par sey oueke les peus ke sunt morz de morine. E kant les berbys serrunt venduz de ceo o de lur leyne e des peaus auandites releuet autant des tetes. Vne gent leuent de ceuz ke sunt mors de morine autres. Coment? io le vos dyray. Si vne berbyz murge sudeynement il mettent la char en ewe autant de hure com est entro mydi e noune e pus le pendent sus e kant le ewe est escule le fount saler e pus ben secher. E si nulle berbyz comence a descheyr ke il veent si ceo ne seyt par la reson ke.les denz lur cheunt. E si les denz ne ly chete mie le fount

---

[1] ne.
[2] [les berbiz le vunt eschwan il a eus.] nest pas bon signe ke il seit debonere a eus.]

How you ought to sort out your swine.

Sort out your swine once a year, and if you find any which is not sound take it away. Do not have boars and sows unless of a good breed. Your other female swine cause to be kept, that they do not farrow; then shall their bacon be worth as much as that of the males. And let them be able to dig. They have need of help in three months, in February, March, and April. Three times a year ought your sows to farrow, unless it be for bad keeping. It is a good thing for swine to lie long in the morning, and to lie dry. Let your sucking pigs be well kept, and they will grow the better.

How you ought to sort out sheep.

See that your shepherd be not hasty, for by an angry man some may be badly overdriven, from which they may perish there where your sheep are pasturing and the shepherd comes among them. Sort out your sheep once a year, between Easter and Whitsuntide, and cause those which are not to be kept to be sheared early and marked apart from the others, and put them in enclosed wood or in other pasture where they can fatten, and about St. John's Day sell them, for then will the flesh of sheep be in season. And the wool of these may be sold by itself with the skins [of those] which died of murrain. And when the sheep are sold, for them and their wool and the skins aforesaid replace as many head. Some men replace others for those which died of murrain. How? I will tell you. If a sheep die suddenly they put the flesh in water for as many hours as are between midday and three o'clock, and then hang it up, and when the water is drained off they salt it and then dry it. And if any sheep begin to fall ill they see if it be because the teeth drop, and if the teeth do not

tuer e saler e sechyr com lautre. E pus le funt perser e
despendre en le hostel entre serianz e ouerours. E atant
cum le pris amunte, rendunt e sesun e de ceo e des peaus
releuent autant des testes mays io ne voyl mye ke vos vset
cest manere. Veet ke vos berbyz seyent en mesun entre
la seynt martyn e pasche io ne dy pas si la tere seyt secke
e la faude seyt atyre a son dreyt e estrame e le tens seyt
bel ke vos motuns i gysent. E ceus ke sunt en mesun ke
yl eyent du feyn ou plus ou meyns solom ceo ke le tens est.
E fetes marler la eyre de la bercherye checune quinzeyne
sicom vos ny dyt auant e ke seyt estrame sure e sachet
vos aueret ayceles plus de pru ke si eus guesent en faude.
E si vos mutuns seyent en mesun pur tempeste ke il seyent
par eus e eyent del plus gros feyn ou le feyn medle oue
forage de furment ou de aueyne ben batu pur quey? io
le vos dyray il sunt debatu la nuyt en la faude par
cheance e lendemeyn ensement ke ne poent pestre e pus
vynent a la crache familous e reboutent les febles e transglutent sanz mascher le feyn menu. E la berbyz kant a
mange sa saule si rounge e ceo ke nest pas masche ne
vent pas a ronge mes demert de denz le cors e purryst
desnaturelment dount plusurs sunt pery. E si le forage
seyt medle il le mascherunt de meus par la grossur del
forage. E si vos auet de faute defcyn le escorces e les corgeys de pesaz est bon as motouns.[1]

Coment hom deyt aygneus garder.

E quant vos aygneaus sunt aygnelez ke le bercher ouste
la leyne entur les treouns kar souent auent ke la leyne
satache as bouches des ayneaus par les treons ele gloutet e
demurt en lur estomak e par tant sunt plusurs periz.

---

[1] More matter follows at this point in the Oxford MSS. see pp. 36, 37.

fall out they cause it to be killed and salted and dried like the others, and then they cut it up and distribute it in the household among the servants and labourers, and they shall then yield as much as they cost, for by this means and with the skins they can replace as many. But I do not wish you to do this. See that your sheep are in houses between Martinmas and Easter, I say not if the weather be dry and the fold be prepared properly and strewed, and if the weather be fine your sheep may lie there, and let those that are in houses have more or less hay, according to the weather. And marl the ground of the sheepfold each fortnight, as I have said before, and let it be strewed on the top, and know you shall have from these more profit than if they lie in the fold. And if wethers be in the house for a storm let them be by themselves, and let them have the coarsest hay or hay mixed with wheat or oat straw, well threshed. Why? I will tell you. They are driven for the night in the fold, and by chance the morrow also, that they cannot pasture, and then come to the manger starving, and push back the weak and choke themselves without chewing the small hay. And when the sheep has eaten its fill it ruminates, and that which is not chewed cannot be chewed again, but remains within its body, and wastes unnaturally, whereby several have perished. And if straw be mixed with the hay they will chew it better because of the coarseness of the straw. And if you have lack of hay the pods and straw of peas are good for sheep.

How you ought to keep lambs.

When your lambs are yeaned let the shepherd take away the wool about the teats, for often it happens that the wool adheres to the mouths of the lambs from the teats, and they swallow it, and it remains in their stomachs, and thereby have many died.

Coment vos deuet vos motuns changer.

A la seynt symon e seyn Iude facet tuer deus de meylurs e deus de myuueyns e deus de pyres e si vos trouet ke eus ne seyent mye scyens fetes vendre vne partye a lele genz par bone surte iekes a la hokeday e donc fetes releuer autres.

Des owes e des gelyns.

Des owes e des gelyns seyt al ordeynement du baylyf e ne mye pur ceo al tens ke io fu baylyf dayes eurent les owes e les gelyns a ferme. Owe a xij d. e gelyne a iij d. e en aucun an valent iiij d.

De vendre en seson.

Vendet e achatet en seson par vewe de vn real [1] homme ou de deus ke pussunt testmoyner les choses car souent auent ke ceus ke rendent aconte en cressent les achaz e amenusent les ventes. Si vos deuet vendre [2] par peys iluk seet auerty kar il ia fraude grant a ceus ke ne se souent entreweyter.

De vewe de aconte.

Veue de aconte fetes ou facet fere par aucun de ki vos fyet vne [3] foys en lan. E final acounte au chef del an. Vewe de aconte fu fet pur sauoyr le estat de la chose come des issues receytes ventes achaz e autres despenses e de surse de deners si la ke il seynt leuez e des meyns des serianz remuez. Car souent auent ke serianz e prouoz par euz ou par autres fount marchandyse des deners lur seynur a lour prou e ne mye al prou lur seynur e ceo nest pas leaute. E si arrerage chete sur la conte final ke

---

[1] loaus. [2] [ou achater.] [3] deus.

How you ought to change your wethers.

At the feast of St. Simon and St. Jude cause two of the best ones, and two of the middling, and two of the worst, to be killed, and if you find that they be not sound sell a part by true men for good security, until Hockday, and then replace them.

Of geese and hens.

Let your geese and hens be under the command of the bailiff. Notwithstanding this, when I was bailiff, dairymaids had the geese and hens to farm, geese at twelve pence and hens at three pence, and in another year four pence.

To sell in season.

Buy and sell in season through the inspection of a true man or two who can witness the business, for often it happens that those who render account increase the purchases and diminish the sales. If you must sell by weight, be careful there, for there is great deceit for those who do not know to be on their guard.

View of account.

Have an inspection of account, or cause it to be made by some one in whom you trust, once a year, and final account at the end of the year. View of account was made to know the state of things as well as the issues, receipts, sales, purchases, and other expenses, and for raising money. If there is any let it be raised and taken from the hands of the servants. For often it happens that servants and provosts by themselves or by others make merchandise with their lord's money to their own profit and not to the profit of their lord, and that is not lawful. And if arrears appear in the final account let them be speedily raised, and

seyt hastiuement leue. E si eus noument treteyne[1] persones ke deyuent le arrerage les nouns pernez a vos. Kar souent auent ke serianz e prouos sunt detturs eus meymes e funt autres detturs ke poyent ne deyuent e ceo funt il pur celer lur desleaute.

Coment serianz e prouoz se deyuent conteyner.

Ceus ke autri chose vnt en garde quatre choses deyuent auoyr par reson. De amer lur seynur e doter le. E quant a preou frere deusent penser ke la chose est lur. E quant a despenses fere dusent penser ke la chose est autri me poy des serianz e prouoz sunt ke ces quatre choses vnt ensemble si com io quyt mes plusurs sunt ke ount guerpi les treys e retenent le quart e le ount besturne hors de sun dreyt curs. Ben seuent ke la chose est autri en ne mye lur e pernent a destre e a senestre par la ou il quydent meux[2] ke lur desleute ne seyt aperceu. Vos choses visitet souent e fetes reuisit car ceus ke treuent par tant escheuuerunt le plus de mal fere e se penerunt de meux fere.

Cy fynyst la dyte de hosebandrie.

[1] certeyne.   [2] [estendre.]

if they name certain persons who owe arrears, take the names, for often it happens that servants and provosts are debtors themselves, and make others debtors whom they can and ought not, and this they do to conceal their disloyalty.

How servants and provosts ought to behave.

Those who have the goods of others in their keeping ought to keep well four things: To love their lord and respect him, and as to making profit, they ought to look on the business as their own, and as to outlays, they ought to think that the business is another's, but there are few servants and provosts who keep these four things altogether, as I think, but there are many who have omitted the three and kept the fourth, and have interpreted that contrary to the right way, knowing well that the business is another's and not theirs, and take right and left where they judge best that their disloyalty will not be perceived. Look into your affairs often, and cause them to be reviewed, for those who serve you will thereby avoid the more to do wrong, and will take pains to do better.

Here ends the treatise of husbandry.

SUPPLEMENT

TO

LE DITE DE HOSEBONDRIE.

From Merton College Library. No. CCCXXI.

Si les denz eent la maladie dez verms eles seent bien arosez e puis les metez en vne meson ben forme ensemble autaunt de houre com de prime si qe a noune issi qe eles se puissent ben entrechaufer e suer e ceo est vne bone medecine pur vne maladie qe est apele en engleis pokkes. Si vous veez au matin qe vne bone rosee com ceo fust vne teine de igraingnes pendant sur le herbe entre le seint martyn e la feste seint Barthelmeu ne lessez pas voz berbiz issir hors de la faude en cele seson ieqes atant qe cele rosee seit nettement abatue. E auant ceo qe eles soient hors isuz de la faude le bercher les face leuer denz la faude auant qe eles issent car si ceo ne seit fet eles getteront lor feens en chiminaunt dehors la faude e ceo ne serroit mie preu e qe mesmes ceus berbiz voisent persaunt pur ceo qe eles vnt geu longement en la faude par la rosee qe tant ad duree. Estre ceo la rosee de vespre est seyne as auant diz berbiz. Estre ceo si vos auez pasture de bruere ou de more e le tens de este seit moistous ostez vos berbiz qe eles ne pessent en cele moisture par qei ieo le vos dirroy. Bien sauez qe le ewe qe demoert en cretine deuient neire ou iaune ou vermaile e ceo sunt ewes qe ne sunt mie seines car si vn cheual enbust parauenture il aueroit la chaude pisse e bon le sachez ore veignent les berbiz qe sunt pessaunt en cele more ou en bruere e lechent de cele mauueise ewe en pessant e demoert cele ewe en lor cors e auient a la fiez qe eles

comencent a eschaufer par la auantdite ewe primes prent colour de blaunc puis de jaune ieqes en purture e le volez vos veer Entur le seint michel tuez vne partie dez berbiz qe sunt en mesme la pasture e vos trouerez qe ieo vos di voir. E si vos volez qe eles soient sauez quant la cretine vient en lauant dite pasture en este tens le ostez e les metez en pasture sekke.

From the Bodleian Library, Digby, 147.

Aliud signum habent bercarii ad cognoscendum, an oues sint corrupti. Accipiunt oues et euertunt palpebram et inspiciunt venas circa oculos et si rubre fuerint, signum est sanitatis; et si albe, signum est infeccionis. Item aliud accipiunt ouem, et in latere post costas diuidunt lanam et temptant an lana firmiter [sic] pelli adhereat; et si ita est, bonum signum sanitatis est; et si lana cito euellatur, signum insanitatis est. Vlterius inspiciunt pellem ouis cum lana fuerit diuisa, et digito calefaciunt pellem, mouendo digitum super pellem, et si pellis deueniat subrubea, signum sanitatis est; si alba vel pallida, signum corrupcionis est. Aliud signum habent bercarii: in principio anni cum gelu venerit circa festum omnium sanctorum, mane cum venerint ad ouile inspiciunt oues, et ille [sic] quarum vellera gelata sunt reputant sanas; non gelatas propter nimiam et innaturalem calorem, reputant infirmas et non sanas. Necessarium est etiam quod bercarii inspiciant pasturam, quum aliquando accidit quod in mane multe albe testudines in pasturis apparent, et cum hoc inspexerint non permittant oues exire a falda, donec calor solis incaluerit, quia tunc repunt in terram. Item vtile est ut sagaciter inspiciant rorem qui vocatur meldeeu quia ille ros inficit oues si fuerit ab eisdem receptus. [Qui ros sic cognoscitur. Qualibet die cum ros fuerit, recipiat pastor virgam coruleam et madefaciat in rore, et faciat rorem distillare per virgam et si gutte adhereant in descendendo sicut seruisia rubea vel viscosa, tunc ros est infectus,

et si gutte descenderint curte non viscose, consideratis supradictis, potest dimittere oues ad pasturam quam mane voluerit.][1] Iam dicam tibi remedium contra corrupcionem ovium: accipem [sic] ouem infectam vel corruptam, et custodi eam a cibo quasi per diem, inclusam in quadam domo: secundo die recipe furfur triticeum, et bonam quantitatem salis et simul misceas et madefac cum aqua, et pone vas cum aqua mixta, et nichil aliud comedat per tres dies continuos et de illo sufficienter habeat [post ponatur ad pasturam et illud anno [sic] viuet].[2] Istud ponit Bartholomeus de proprietatibus rerum[3] in testimonium cujus audiui a fidelibus ouem matricem durasse in bono statu in salcis pasturis per xxxiij annos, et certum est quod sal fuit in causa preservant et eam desiccant.

[1] This passage is written on the margin. in the printed copies of the *De Proprietatibus Rerum* of Bartholomew Anglicus, sometimes erroneously called Bartholomew De Glanville.
[2] These words are inserted between the lines.
[3] The reference does not appear

TRANSLATION

OF

WALTER OF HENLEY'S HUSBANDRY

ATTRIBUTED TO

ROBERT GROSSETESTE

TRETYCE OFF HOUSBANDRY

Thᴇ tretyce off housbandry þᵗ maystur groshe?[1] made þe whiche was bishope off lycõll he transelate þis booke ovt off ffrenshe in to englyshe þe begyninge off þis booke techithe all maner men ffor to goũne þ́ landis tenementꝭ and demaynes & ordynately to rule þ́ begyninge the chapituris and þe table acordynge off þe same booke.

The firste chapitur tellithe howe ye shall spende yoᷓ good and howe þe shall extende yoᷓ londis.

The secunde chapitur tellithe howe yoᷓ land shall be mesured and howe many perchis of londe makithe an acre and howe many acres makithe an yerde off land and howe many yerdis makithe an hyde of land & hou many hydis makithe a knyghtꝭ fee.

The thirde chapitur tellithe howe many acres off land þat a ploughe may tyll in a yere.

The iiij chapitur tellithe wheder a ploughe off oxon or a ploughe off hors may tyll more land a yere & whiche of þem is more costfull.

The v chapitur tellithe in what seasone ye shall begynne to falowe yoᷓ lande in all mañ landis.

The vj chapitur tellithe nowe howe you shall lay youre lande at seede tyme.

The vij chapitur tellithe howe yoᷓ lande shall be sowen in all seasons.

The viij chapitur tellithe howe ye shall change your seede and norishe your stoble.

The ix chapitur tellithe howe you shall norishe yoᷓ doung & wede yoᷓ corne and howe it shall be mesured ovte off yoᷓ berne & howe moche a acre off land shall yeld agayne more þen yoᷓ seede yeff ye shuld haue wynyng þ́ bye.

[1] MS. torn.

The x chapitur tellithe howe ye shall change all maner off catell in seasone.

The xi chapitur tellithe howe ye shall norishe your werkyn beestę and wayne yoᵘ calvis & what approumentę ye shall haue off your kyne & ewene in chese and butur.

The xij chapitur tellithe howe ye shall norishe yoᵘ swynne and yoᵘ peggę.

The xiij chapitur tellithe off norishynge off shepe and off dyuerse medsyons for them.

The xiiij chapitur tellithe what enpproumētę ye shall haue off yoᵘ gesse and hennys.

The xv chapitur tellithe howe ye shall by and sell and preve your weyght in all seasons off the yere.

The xvj chapitur tellithe off acompte & off avewe off yoᵘ baylis throughe of þe yere.

The xvij chapitur tellithe howe ye shall graffe & plante all treis & vynys off all frutę.[1]

The firste chapitur.

The ffader in his old age seithe to his sonne leve wisely and discretly aftur god & þe world & thynke on the harde change off ffortune howe by lytell & lytell yt attaynyth to richis & by lytell it discendithe to pouerte & aftur in to myche vnease or wrechenys wherefore i cownsell you to ordeyne yoᵘ lyuinge aftur the extente off yoᵘ lyvelode & not more þen ye may dispende in a yere by yoᵘ lyvelod and yeff ye may approwe and make yoᵘ londis betł by wyning or ellis by store off catell or eny oþer approuynge more þen thextente off yoᵘ lyvelode amonuthe as moche þen it is more in valewe thextente kept & dispende it not for yeff yoᵘ store off catell dye yeff yoᵘ corne fayle it may stonde you in good stede for yef you dispende the value off yoᵘ lyvelode in a yere & mysaventure fall vpon you ye haue no rekeū by yoᵘ approwmentę for þe wiseman seithe he þᵗ approvith to oþ men often tymis he wastithe his owne for it is seyne þat many men haue bothe londis & tenementę & can not leve þ vpon for enchesone þᵗ þey lyve wᵗ ovt

---

[1] But the treatise ends with the sixteenth chapter.

ordenᵃūce & purvyaunce made in due seasons & for þey
haue spent more þen p̄ lyvelod may suffice & menteyne &
þen þey can none oþer shifte but fro þe hande to þe mowthe
& soo þey fall in poūte & wrechidnes þer for be wise in yoᶻ
demenynge desire none off yoᶻ neghbours good wrongfuly
and kepe yoᶻ owne wisely so þᵗ by reasone no man may
repreve you nother your werkis þe good þᵗ god hathe lente
you dispose it wysely in yeftis and dispensis ye shall see
iiij þyngℓ þat shall be profetable to you. The firste is when
ye geve and to whom. The secunde is yeff ye shall geve
or make eny dispensis loke it be done withe good will and
þen it shall be preysed ffor yeff yt ayenste yoᶻ wyll ye lesse
as moche as ye yeve or dispende. The thirde is loke ye
geve to hym þᵗ may bothe forder you & hynder you. The
iiij is howe ye geve noþer to moche ne to lytell but aftur
þe persons by þᵗ ye yeve to aftur your mater be mykyll or
lytell and aftur þᵗ ye haue to doo wᵗ þe persones let yoᶻ
landis be extendide by wyse men sworne and se what eūy
parcell p̄ off may be worthe in a yere more þen thextente
as wele in yoᶻ manor as in yoᶻ gardynes doofe houses and
closis and afturwardis loke howe many acres be in yoᶻ
closis & se what a acre p̄ off is worth by yere see also howe
many acres off erybyll londe ye haue and what acre p̄ off
is worthe by yere and off all oþ londis medewis pasturis in
lyke wyse and here by shall ye knowe what ye may dispende
in a yere wᵗ ovt waste or distrucyone and of fre tenutres
acordynge howe moche eūy man holdithe off you and by
what servyse also what myls or fishyngis be worthe yeff ye
may haue withe in your lordshipis yerely oū thextente and
in semable wise off all mañ cotagℓ rentℓ fermes costomes
& all oþ þyngℓ whereby eny profet to you yerly may aryse
and loke howe moche seede shulde competentely sowe a
acre off lond off eche maner off graynes in his kynde and
loke howe moche store ye may haue wpon eūyche off yoᶻ
maneris & þis knowe þan se thextente off yoᶻ lyvelode and
ye shall redely discerne off howe moche yoᶻ baylyfe shall
answere you yerely by þe approwment oū thextente & in
þis shall ye vnderstonde veryly what is þe yerely value of

yoᶻ lyvelode & þ̄ vpon rule your dispencys prudently as
j haue cownseled you before also loke ye take a redy
rekenynge off yoᶻ bayle yerely in þe monethe off juȳn howe
many quarteris off all mañ off corne is sowen vpon yoᶻ
demayn londis and þen loke your extente & goo in to yoᶻ
feldes þ́ wᵗ & take a vewe off eūy pece off londe þat is
sowen & paraventur ye shall fynde more corne & more
londe sowen þan he gevithe you acompte off þe whiche he
wold kepe counsell & kepe prevely to his pper vse behove
& advayle or elis paventur more corne sowen vpon yoᶻ
lande þen nedithe & þis shall ye knowe for very sertente
wheder yoᶻ bayle be good & profetable for you or no and
yeff it lyke you ye may depart yoᶻ londis in iij partis The
firste parte to be soven wᵗ wyntur corne þe secunde parte
to be sowen withe lenten corne as with otys pecys barly &
soyche oþ́ graynes The thirde parte to be falowed & somer
layd & yeff ye wyll ye may departe yoᶻ lande in ij partis
þᵗ one þ́ off myght be sowen with wynter seede & lenten
seede þᵗ oþ́ parte to be falowed & som̄ layde & þis shall a
ploughe wele tyll viijˣˣ acres off lande in a yere þ́ fore loke
yoᶻ extente & se howe many acres off lande ye haue and
comaunde yoᶻ bayle straytly to kepe þis mañ off gydynge
in telthe.

The secunde chapitur.

It is to wite þat iij barley cornys þat is in þe mydiste
off þe eyre makithe a enche and xii enchis makithe a foote
& xvj fote and a halfe makithe a pche & fourti pchis in
lengthe & foure in brede makithe a acre off londe & iiij
acres makithe a yerde of londe and v yerdis makithe a
hyde off lande & viij hydis makythe a knyghtę fee.

The thirde chapitur.

Som men seyne þᵗ a ploughe may not tyll nor susteyne
viijˣˣ acres or ixˣˣ acres of lond in a yere but i shall preve
you by good resone that a ploughe may do yt for ye shall
vnderstonde þᵗ a acre off londe is in mesure xl perchis in

lengthe & iiij in brede and the mesure off a perche is xvj
fote and so þe brede off a acre off londe is xlvj [sic] fote so go
withe youre ploughe xxxiij tymis vpe & downe þe lande and
se þat þe firste foroughe be a fote & all þe toder forowis off
lyke quantyte & þᵗ is a acre ered and when þe forough is
as straythe as it may be þen it is xxxvj tymis vpe & downe
þe lande with þe ploughe thoughe it be a large acree & þe
ploughe be neū so feble þen yet at þe moste ye haue gō
but lxxij vpe and down þe lande & þᵗ is but v myle way
nowe trewely þe ox or þe hors is ryght febyll þat fro þe
morowe may not go softely iij myle fro his home and com
agayne be noone and by þis oþer reasone ensuynge j shall
shewe you þat it may be do as moche ye knowe wele þat
p̄ erne in þe yere lij wekis & viij wekis for holydays and
oþer letyngꝭ and yet p̄ levithe behynde xliiij wekis to werke
in these xliiij wekis be xiijˣˣ & iiij dayis be syde sondays &
it is to wite þat a ploughe shall erye iij tymis in a yere þᵗ
is to say in wyntur in lentyn and in lyke seede tyme in
wyntur a ploughe may erye iij rodis & a halfe on þe day &
in eche off þe oþ̄ seasons a acree off þe day at þe leste nowe
knowe ye wheder it may be do or nay but by cawse þat
ploughe men carteris and oþer seruantꝭ feyne & werke not
trewly it is behoue full þat men ffynde a remedy ayenste
ther seruntꝭ where fore it is neccessary þat þe bayle or som
off þe lordis offeceris be withe þem þe firste day off erynge
falowinge & sowynge to see þᵗ þey do trewly þer werkis &
aftur þᵗ let þem answere dayly off as moche werke as þey
did on þe firste day as when þey were oū sayne but yeff þey
can fynde a resonable excuse off ᴣher¹ distorbance also it
is neccessary þat your bayle oū se yoᶻ werke men onys a
day to wite yeff þey do p̄ werke soffecyently as þey ought
to do and he fynde þem þe contrarye he for to chastice þem
resonable p̄ fore.

The iiij chapitur.

The ploughe off oxon is bettᵽ þen the ploughe off hors
but yeff it be vpon stony grounde whiche grevithe sore the

¹ erasure.

oxon on þͤ fote for þe ploughe off hors is more costfull þen is þe ploughe off oxon & yet shall þe ploughe off oxon do as moche werke in a yere as þe ploughe off hors for thoughe yeff ye dryve yoᷓ ploughe off hors faster þen yoᷓ ploughe of oxon yet on what ground so it be yoᷓ ploughe off oxon yef ye tyll yoᷓ lande wele & evynly þey shall do as moch oo day wythe another as yoᷓ ploughe off hors & yeff þe grounde be thoughe the oxon shall werke þͤ where as þe hors shall stonde styll & yeff ye will wit howe moche þᵗ þe tone is costilychere þen þe toþͤ is i shall tell you it is costom þᵗ beestͤ whiche go to ploughe shall werke fro þe feeste off seynt luke vnto þe feeste of seynt elyne of holy-roode in may þᵗ is to say xxv wekͤ & yeff þe hors shuld be kept in good plyght to do his jorney he moste haue dayly at þe leste þe sext parte off a bushell off otys þe pryse a ob. in grasse in somͥ seasone xij d. at the leste & eûy weke þᵗ he stondith to drye mete one withe another oƀ. in strawe for letͭ & in shoynge as often tymis as he is shode on all iiij fete iiij d. at þe leste þe som of his expensys in þe yere is x s. & v d. ob. besydis haye & chaffe & oþͤ þyngͤ & as for þe oxon ye may kepe hym in good plyght dayly for to do his jorney yevynge þem eûy weke in ote shevis þe prise a j d. for be cawse þat x ote shevis yeldyn a bushell of otys yeff þey be made by thextente & it be in somer seasone he moste haue xij d. in grasse the hole som here off is off expensis in a yere iij s. j d. be sydis strawe & chaffe where as a hors is wered and oû set and brought downe by labur it is to aventur yeff he oû˙ rekeû it and thoughe yoᷓ ox by labored & be so wered & oû set and brought downe ye shall for xij d. in somer haue hym so pastured þᵗ he shall be strong inowghe ayen to do yoᷓ werke yeff ye will or ellis he shall be so fate þᵗ ye may full fayne sell hym ffor as moche money as he coste hym.

The v chapitur.

In aprell it is good seasone to falowe land yeff it be broke wele afore þe ploughe ffor in þᵗ seasone it is nother to wite ne to drye but he þᵗ hathe moche to do may not

abyde þe good season off þe yere neū þe lesse when so eū
þey erye yeff it be good soyle eree depe wᵗ a square forough
so þᵗ som off þe good lande may reste & yeff þe lande be
nother keūed ne vnkeūed at þe secunde falowe ereye not
depe but so as ye may stroye þe thistelis & other wedis for
yeff yoᶻ land ly in marres or in watry grounde & it be
ereyd to depe at þe secunde falowe when ye erye to seede
yoᶻ ploughe shall com to no harde grounde but go schoutyng
all in myrre and yeff yoᶻ ploughe go a enche deppere in
seede tyme þen it did at the secunde falowe it shall fynde
good grounde & clense þe telthe wele off þe myrre & make
good eryeing & clene.

The vj chapitur.

At sowing off yoᶻ seede lay yoᶻ lande narowe inowghe
togeder so þat yoᶻ sede may fall evynly on yoᶻ lande for
shortely to declare yeff ye lay yoᶻ lande wyde asonder ye
shall do youꝛselfe¹ grete hurte for when þe lande is sowen
þe harowe shall caste þe corne in þe hoolis & valeis þᵗ
betwixt þe gatas off þe ploughe so þᵗ þe corne þᵗ is in þe
ryge off þe lande shall be vnkeūed by cawse where off lytell
or nowght shall growe vpon yoᶻ lande & yeff ye will prove
it when yoᶻ corne is growen ovt off þe erthe go to þe hede
off yoᶻ lande & loke towarde þe toþ end and þen shall ye
see wheþ j say trewe or no and yeff sowe yoᶻ lande vnd
þe foroughe let it be ereyd & layd small and neyghe togeder
so þᵗ þe ryge off þe londe betwene off þe foroughes by
narowe inowghe togeder like a creste in þe mydis off þe
lond ryge vnder þe lefte foote & when þey ere þe same lande
ayeyne kerue yt withe youre ploughe so that þe firste
erynge may be oū turned & þan shall yoᶻ lande lye neyght
inowghe togeder.

The vij chapitur.

Sowe yoᶻ wyntur corne tymely so þᵗ your lande may be
sadid & yoᶻ corne rotyd afore þᵗ grete wyntur com for yeff
a grete rayne ffall withe in xx dayis aftur þe sowing off

---

¹ erasure.

your corne & þ vpon a froste enduryng ij or iij dayis þe
froste shall make þe corne to pishe þᵗ is nowe sowen for
enchesone þᵗ it is but tender and þe rote þ off is but newly
budid or put ovte all soyche clay londis & stony londis as
ye purpose to sowe withe lynten seede sowe þem tymly
afore þᵗ may com for may makithe þe clay londe herde &
þe stony londe drye so þ fore cōmonlyke þe londe openyte
for drynesse & þ fore it is nedfull to sowe soyche mañ off
londis tymely so þᵗ þe seede þᵗ is sowen þer vpon may haue
his norishynge & rotynge as ffor sond⁾ londis it nedid not
to sowe þem so tymely for it is not good to plowen soyche
mañ o londis in grete moysture yet thoughe þey be alitell
wete withe a dewe at þe soweinge it shall not ney þem but
do þem moche good and yeff your land ly in marysse or in
watry grounde make good depe foroughis þ in so þᵗ þe regẹ
may be delyůed off þe watur.

The viij chapitur.

Chaunge yoᶻ seede eury yere myghelmas for it shall be
more advayle for you to seede yoᶻ londes withe seed þᵗ
growe on oþe mennes londis þen withe seede þᵗ growe on
yoᶻ owne londẹ & yeff ye will make a preffe herin erye
too pertis off londe lyke in luste & lyke in soyle & sowe
hym all at onys þe one withe seede þat growe vpon yoᶻ
owne londe and þat oþ with seede þᵗ growe vpon oþ menys
londis and þer shall ye see wheder j say sother or no take
not stoble off yoᶻ londe þ as it growe lesse þen ye haue
grete nede þ off ffor couerynge off yoᶻ howses.

The ix chapitur.

Take yoᶻ dongge & medell it withe erthe þᵗ is freshe &
make clene yoᶻ shepcote at eury xv days ende and medel
þe doung þᵗ comythe þer off withe freshe erthe or elis withe
claye or soyche mater as ye caste ovte of dykis & strawe it
wele withe strawe & chaffe & yeff ye haue more strawe þen
yoᶻ store shall spende strawe it in yoᶻ folde or in yoᶻ
shipcote & in þe tyme let all þe doung as welle þe whiche
is withe ovte þe shipcote as þᵗ whiche is withe in let it be

gadred togedere and layde vpon a hepe & when ye shall
carye it & lay it on yo⁼ lande take hede þat yo⁼ carteris do
trewly þer werke & let þem answere you dayly off as moche
werke as þey did on þe firste day when þey were oũ sayne
lesse þem þey haue a resonable excuse and cawse off letyng
& when ye dylyũ yo⁼ carteris hors shone sadelis colouris
or oþ̃ harnes loke þᵗ þey delyũ you all þᵗ is brokyn or
apayred or elis acompte it on þ̃ wagis yeff it be loste
throughe þer neglygence and wᵗ þat grasse þᵗ ye dylyũe to
yo⁼ carteris medell it wᵗ harde sope or tarre and it shall
be betur for yo⁼ carte & yeff yo⁼ carteris gresse þ̃ shone þ̃
withe it shall brenne and rote þem to dung þe dounge þᵗ is
medled with erthe let it be put vpon sondy grounde yeff ye
haue eny for in soffi þe wheder is hote and þe sond is hote
be kynde whiche ij hetis when þey mete aftur mydsoffi it
do þe corne to wex passyng sore growithe vpon soiche
sondy londis but yeff to be remedid by þis mañ of doungyng
as it is afore seide when ye erye dunge in the grounde to
sowe corne erye not to depe for that waistithe yo⁼ dounge.
Nowe shall i tell you what wyning ye shall haue by yo⁼
dounge þᵗ is medled withe erthe þe doung þat is rotyn by
hym selfe wᵗ out erth for it shall laste ij yere or iij leñg
and yet aftur þe lond þe hote or cold and dunge þᵗ is
medeled withe erthe shall leste double as moche tyme but
it shall not be so sharpe ne so ranke beryng ye shall knowe
þat marle lastithe leñge þen dunge for dung wastithe &
discendithe & marle mountithe & ascendithe dung medled
with erthe lastithe leñg þen dung not medled withe erthe
for be cawse when dung & erthe medled togeder is be spred
vpon the londe & þe lond harowed þe erthe kepithe þe
dunge þat it may not waste in discendinge as it wold elis
do naturaly and þerfor soyche dunge is beste & moste
profetable & yeff it rayne a litell when ye lay yo⁼ doung on
yo⁼ londe it doithe moche good for it cawsithe þe dunge &
þe lond to joyne well togeder yeff þat you put yo⁼ dunge
vpon your londe at falowyng tyme at þe secunde falowe it
shall be turnyd vnder þe erthe so þᵗ it shall be clene vnder
þe erthe at sowyng tyme & þᵗ is not gretiste profete let þe

E

dunge off yo⁼ shepe be put next yo⁼ sede ffor so it is
mooste worthy at þe firste feste off oure lady ordeyngne
hidles aftur þᵗ ye haue shepe & rere yo⁼ folde for in þᵗ
seasone þey haue mooste dounge & aftur mydsom̅ let yo⁼
corne be wedid and made clene and not affore ffor yeff ye
kyt thistelis xx or xv or viij dayis afore þᵗ tyme for euy one
thistill iij or iiij shall growe ayeyne let yo⁼ medowis be
wele and clene mowen by þe advyce and vewe off yo⁼ bayle
and se welle þᵗ yo⁼ mowere hold not his ryght honde afore
to hyghe be hynde hym so þat he kyt a sonder þe grasse
in þe mydis and þis defaute is callid forsyng & many men
take lytell hede p̔ to & it is a grete defaute & a grete losse
off hay let yo⁼ corne be wysely shorne and gedrid & laied
in yo⁼ berne and let yo⁼ thresers be sworne to thresse it
clene neu̅ þe lesse take heede to þem þat þey haue no
poketẹ nor grete purses where as þey myght stelle & bere
away your corne also se þat yo⁼ wenyheris haue no poketẹ
betwene p̔ leggẹ to stelle withe yo⁼ corne shall be take out
off yo⁼ berne purvey you off a trewe man ffor to ou̅ se yo⁼
bayly for it is often seyne þᵗ all þe offecers bene off one
asente to avayle þem selfe and hurt þe lorde se also þᵗ yo⁼
corne be mesured withe a trewe mesure that is to say a
trewe bushell & þᵗ euy bushell be strekyn and se þat þe
mesures haue a clothe vnder þer fete to kepe þe corne þat
falithe when it is put in to þe sekkis also be wele ware
off mesurynge off yo⁼ bushell þᵗ is vphepide for p̔ in is
grete disceythe & i shall tell you howe when þe bayle hath
made his acompte off þe corne let þe bushell where by ye
reseyuid þe corne be previd with his accompte ffor yeff it
be a large bushel iiij bushelis vphepid makithe v strekyn
or lytel more or lytell lesse or elis v bushelis vphepid
makithe vij bushelis strekyn yeff þe bushell be not so large
or som what more & yeff þe bushell be neu̅ so smale vj
bushelis vphepid makyn vij strekyn & so euy bushell som
tyme more and som tyme lesse p̔ be som baylis þat yevyn
at p̔ accompte for viij bushelis vphepid but ix bushelis
strekid and wheder þe bushell be lytell or mykyll p̔ in is
grete disceyte for at þe grete bushell he stelithe ij bushelis

& yeff þe corne be grete & large þen bothe at þe grete
bushell and at þe smale bushellis both falsenys & grete
disceite yeff yoᵣ londe yeld agayne but thre tymis as moche
corne as ye sowen þ̃ upon ye shall wynne noþinge þ̃ by but
yeff corne hapyn to be at a gret̃ prise þen it was when it
was sowe for ye shall vnderstonde þᵗ a acre off londe shall
haue iij erthis or þan it be sowen & som more and eũyche
off these erthis is worthe vj d. þe harowyng a j d. & of the
seid acre off londe shall be sowe ij bushelis wete þe price
xij d. & þe wedynge þ̃ off ob. þe sheryng v d. þe ledyng in to
þe berne j d. and the strawe & þe chaffe shall aquite þe
threshyng & so iij tymis þe seede off the same londe is vj
bushelis & yeff a quartur off whete be worthe but iiij s.
aftur myghelmas þen yoᵣ vj bushelis are worthe iij s. & yoᵣ
costis done vpon þe seid acre drawithe iij & j d. ob. be syde
þe rente off þe ferme þᵗ þe lorde hathe þ̃ fore.

The x chapitur.

And yeff ye haue eny londe where vpon is store off
catell may be norished & kepte let it be stored aftur þᵗ it
may bere for yeff it be wele stored & þe store well kepte &
tendid it shall answere you off as moche as thextent off
yoᵣ lande amountithe but loke þᵗ ye drawe and serche yoᵣ
store off all mañ catell onys in þe yere and betwene ester &
withsonday & change those þᵗ be not good to kepe & þᵗ as
wele off yoᵣ carte hors & oxon as off all oþ̃ catell and those
þᵗ be not goode to kepe put to grasse for yeff ye make þem
fate withe grasse ye shall haue wynynge þ̃ by and ye shall
withe þᵗ þe feble ox costithe as moche and more þen þe
beste ox for yeff he be a wayster ox he moste be þe more
spared and by þᵗ sparynge þe best ox is þe more greuyd
& yeff ye shall by yoᵣ store off catell by it betwene estere
and withsontid for þen beestes ben lene & good chepe lok
þᵗ you chaung yoᵣ carte hors or þan þey be sore worne or
spent ffor withe as lytell coste shall ye brynge vpe yeng
hors as for to kepe styll þ̃ old and yeff ye by and sell in
seasone it shall avayle you more þen for to kepe styll yoᵣ
old tyll þey be worne.

The xi chapitur.

Geve yoᵘ ploughe beestꝭ sufficyaunt mete for to susteyne wᵗ þ labur so þᵗ þey be not oūcharged ne oū moche brought down withe labur for yeff þey be oū set it shall coste you to moche or þey be rekeūed ayeyne & relevyd and also yoᵘ werke shall be letyd þ by put not yoᵘ bestꝭ in howses in rayne wheder nor in grete hete for þat engenderithe a hete betwene þe skynne and þe fleshe & betwene þe leske and the thye whiche turnythe yoᵘ catell to grete hurte & yeff yoᵘ catell haue eūy day provender let it be highe day or þey haue it & þen yet by þe dylyūance off yoᵘ bayles & let yoᵘ provynder be medled wᵗ whete chafe or ote chaffe but not with barle chafe for þᵗ hurtithe þem in þe mowthe & specyaly hors ye shall wite whye that provender is medled withe chafe þᵗ the kepers off þe hors stelle it not away and also þe chafe cawsith þe hors to hete and drynke betꝭ þen þey shall elis do & to be fatere þen þey shuld elis be also loke þᵗ þ stable be eūy day made clene ffor þᵗ doythe hym moche good & loke you geue yoᵘ oxon no grete quantatye off strawe at onys but be lytell & often tymis þat doythe hym ete well and waiste but lytell for when þey haue a grete quantatye afore þem at onys þey ete þer fell & þen þey chewe there cude and blowithe on þat oþ mete whiche is lefte so þᵗ it wexithe drye and þen þey wyll no more ete þ off & also loke þᵗ yoᵘ oxon be dayly made clene & robyd withe a wispe off strawe & þen þey shall lyke þem selfe þe betꝭ let yoᵘ kyne haue sufficyante mete and let þem haue þe provender þᵗ your oxon & hors levyn oū nyght & yeff a male calfe be seke when it is calvid let it haue þe moderis mylke a monethe and it at þe monethis ende take fro it a pape & so at eūy wekis ende folowing a pape tyll he haue soukyd firste & loste vij wekꝭ & in þᵗ mone seasone lay mete afore hym and lerne hym to ete let þe femalis calvis haue þe modris mylke iij wekis and at þe iij wekis ende take from here a pape & so furthe wekely as ye did off þe male calve & let yoᵘ calvis haue watur inowe &

let hym not com ovte off þe howses tyll þey be wayned & som what stife off age for many calvis dye for defaute off howses off a evele þᵗ is calyd la pomelyere & yeff eny off yoᵘ beestɇ begynne to ffall in sekenys loke ye spende a j d. betymes to rule and to help it for the wiseman seithe blessed be þe j d. þᵗ savithe twayne & yeff eny off yoᵘ bestis dye in moreyne let here be flayne and put þe skynne in watur viij or ix dayis & þen take it vpe & þen let þe watur renne ovte þer off and þis shall make þe skyn thike & bettɇ to þe sayle but when ye shall sell it let not it be to dry but fow what moyste yeff ye will knowe þe pfite & issue off yoᵘ skynne nowe ye knowe it and off yoᵘ ewene in butur & chese & howe moche a cowe shall eũy weke geve ye moste but þe feble skynne fro þe good kyne & yoᵘ good kyne go in good pasture off salt maries þen ij kyne shall answere a peyse off chese betwixt estɇ & myghelmas & be sydis þᵗ eũy weke off a galone off buttɇ & yeff it be in freshe pastur þᵗ is to say wood or felde or in stoble aftur mawynge þen iij kyne. shall answere off as moche as two þᵗ be pastured in þe salt mares & but off lytell more & as for shepe yeff þey be pastured in salt marris þen shall xx moder shepe answere off as moche as þe to kyne þᵗ gone in salt pastur & yeff þey go in freshe pasture þen shall xxx modris shepe shall answere off as moche as þe thre kyne þᵗ go in freshe pastur þ̕ be many baylis & dayis þat will say nay to þis and but yeff þe mylke be spyllyd or spent oþ̕ wais þey will yeve well so moche withe ovten fayle and i shall tell you howe off þis iij kyne that geve ayeyne a peyse off chese betwen ester & myghelmas and eũy weke a galon off buttɇ þᵗ is a feble cowe þᵗ in ij dayis gevithe not as moche mylke as wyll make a chese off a ob. þᵗ is in vj dayis j d. ob. & þe sondayis is not rekenyd for it is savid for what so eũ necessite þat falite or for þe tethe ye knowe wele þᵗ betwixte estere and myghelmas erne xxiiij wekis & for eũy off these wekɇ rekenyd j d. ob. where off þe somer is iij s. þen put to as moche for þe secunde cowe & for þe thirde cowe & þe som off all is ix s. & so shall ye haue a peyse off chese for þᵗ is þe comyn pryse þ̕ off also þᵗ cow is

ryght feble þᵗ may not geve eũy weke wᵗ þe thirde parte off a potell off butƚ and yeff a galon off butƚ be sold for vj d. þen is þe thirde pte off a potell worth j d.

The xij chapitur.

Loke yoᶻ swynne be drawen & lokyd onys in þe yere anone aftur estere and let thos be chaungyd þᵗ be not hole & loke ye kepe no borys with ovte þey be com off good kynde also loke you kepe yoᶻ female swyne at farowynge tyme so þᵗ þey be not hurte nor apayred ffor defaute off good kepynge & þen withe good maystyng afturward þey shall bo good for yoᶻ lardere as youre smale swyne in wyntur geve yoᶻ swyne mete inowghe so þᵗ þey may haue pouer & be stronge off hym selfe and spesyaly in feuerere marche and aprell ffor þᵗ tyme haue þey moste nede & þᵗ tyme shall yoᶻ sowis haue pyggis withe ovt it before had for bad kepynge & yeff ye will norishe hym wele kepe þem long at þe morowe withe and let þem lye drye while þe piggis sowkyn and þey shall growe þe better.

The xiij chapitur.

Loke þᵗ yoᶻ sheperde be not irous withe yoᶻ shepe for þᵗ is a evill vice and ye shall preve þᵗ þ̄ as þe shepe pasturyne and your sheparde goythe among þem for yeff þe shepe þen fle fro hym it is a sygne þᵗ he is not peseable with þem & loke þᵗ ye eũy yere onys betwixt ester and whitsonday drawe yoᶻ shepe and loke yeff þey be clene withe ovte sekenys & yeff eny of þem be defaute let hym soone be clepid & marked & put away fro þe hole shepe in to a good pasture for to be made fat & at mydsoɱ when þey be fat sell þem for þan is moton in seasone and let þe woll off þis shepe be sold with these skynnes of þem þat dyed in moreyne & withe þᵗ mañ wol & skynnes off þem ye shall bye as many yenge shepe þer fore se þat yoᶻ shepe be in howses betwen seynt martyns day & holy roode day in may neũ þe lesse yeff þe wheder be dry and your fold wele strewed withe strawe i say not nay but þen yoᶻ motons may lye in þe folde & those shepe þᵗ lye in þe howse moste

haue hay aftur þe wheder be som tyme more and som tyme
lesse and loke þat þe howse be close so that no eyre may
com to þ crethe & eùy nyght put newe strawe vnder þem &
se þᵗ þ howse be made clene eùy fortunyght onys and so
shall ye haue the more dunge and more profete off þem as
yeff þey lay in yoᵣ folde and yeff yoᵣ motons lye in howse
for tempaste loke þat þey ly by þem selfe & not withe none
oþ shepe and let þem haue of þe gretiste haye þat ye can
fynde & elis let þ hay be meled withe whete strawe or elis
with ote strawe & i shall tell you whye by cawse off þe
tempaste & evill wheder þᵗ þey myght not fede when þey
were vpon þ pasture so þat when þey com in to þe howse
þey are so hungered þᵗ þe strong shepe put þe ffeble fro þ
mete and þen for þis cawse þey swalowen þ mete all holle
withe ovte chewynge & specyaly þe smale haye & when þe
shepe hathe etyn þ fyll he lythe downe and chewith his
cude & þen þe mete þᵗ is not chewed comythe not vpe with
þat at is chewed but it lythe styll in þe shepis body tyll it
be rotyn by his owne kynde & þ off many shepe sone be
pished & yeff yoᵣ haye be meled withe strawe þey shall
chewe it þe bettᵣ by cawse off þe gretenys off þe strawe and
yeff ye haue defaute off strawe or off grete hay [1] togeder for
it is good mete for shepe and yeff þe evill fall amonge þem
þat is callid þe werwes sodenly let þem be sprenkled wele
with watur and þen put þem in a howse as long whylle as
it is from þe morowe to noone & let þem chause wele on
shepe with anoþ for it is good medsine for þe sekenys and
it is for þe evill þᵗ is callid pokkes & yeff ye se at morowe
a dewe vpon þe grounde that is callid webe off arayne
hongynge vpon þe grasse betwixt þe firste feeste off oure
lady & þe feeste off seynt martyne let not youre shepe ovt
off þe fold tyll þat þe vnholsom dewe be clene gone off þe
grounde and let yoᵣ sheperd do yoᵣ shepe arise and stond
vpon þer fete a good while or þey go ovte off þe folde & aftur-
warde dryve þem in to þ pasture & for as moche as þey were
longe kept in at morowe by cawse off the vnwholsome dewe
let þem pasture lenger at evyn when þe sterris be on þe

[1] [take lynge and haye medled], Wynkyn de Worde's edition.

skye for the evyn dewe is holsome & shall do no shepe noo
harme & yeff ye haue pasture off more or elis off hethe yeff
it be drye soṁ take away yo⁼ shepe fro þat pasture ffor in
dry wheder þe stondyng watur whiche is in soyche mañ off
pasture wexithe oƥ blake grene or yelowe & these wateris
ben noþing holsom for shepe nor for oƥ beestẹ for yeff yo⁼
hors drynke ƥ off he shall haue a evyll callid þe caude pis
& yeff þe shepe drinke ƥ off it abidithe in ƥ bodis so long þᵗ
it makithe ƥ fleshe to be corrupt for firste it cawsithe þe
fleshe for to wex withe and aftur þᵗ yellowe & þen sone
aftur þey shall rotyn wᵗ ovte remedy and ffor verey
prowe here off take som off þe shepe þat gone in soich
pastur abovte myghelmas tyme and se þem & ye shall
fynde it þat i say trewe yeff ye will save yo⁼ shepe in a
wete soṁ take þem ovte off wete pasture and put þem in
dry pasture and at þe feeste off seynt Symon and Iude sle
thre off yo⁼ beste wedris & two off yo⁼ beste ewen and yeff
ye thynke þᵗ þey be defaute let þem be sold & rather þen
ye shuld fayle off chepmen sell þem to festẹ so þat ye may
be sure off yo⁼ money þen geder these erbis vnd) writen in
herueste þat is to say ameroch oþer wise callid maydens
wede & dry it & at þe firste comynge in to þe howse of yo⁼
shepe þat is to say at martymas let þis erbe be medled
withe ƥ haye and put som withe in þe walis off þe howse
for it will drye þe evyll humore þat is withe in ƥ bodis and
it is good for þe lyuer and also it distroythe þe wenne wᵗ in
ƥ bodis when ƥ lammes ben ewid let yo⁼ sheperde take
away þe woll abowt þer modris papis for it happithe often
tyme þat it falithe abowte þe lambis tethe and strangelith
þem and som go in to ƥ bodis & cawsith þem to perishe
and let the sheperde answer you off the woll that he
gederithe in þis mañ wise & let it be put to yo⁼ flees when
yo⁼ shepe are shorne and loke þat ye make taylis betwixt
yo⁼ sheperd and your tayle the wedris by þem selfe and þe
ewis be þem selfe þe male hoges by þem selfe þe female
hoggẹ by hym selfe & þe lambis by þem selfe & let yo⁼
sheperde answere off eche parte by þem selfe ij tymis in þe
yere þᵗ is to say at þe feeste off seynt martyne when þey

com in to the howse & at þe feeste off holy roode tyde or
lesse at clypyng tyme when so eũ it be & when ye put þem
in to þe howse let þem be marked þe ere & ordygne you a
strange iron to marke þem withe in þe forhede & yeff eny
off þem dy in moreyne reseyue not yo⁼ skynnes withe ovte
þey be marked withe yo⁼ marke.

The xiiij chapitur.

Geese and hennes shall be at þe orden*ūce off yo⁼ bayle
or elis þey shall be let to ferme a goose ffor xij d. in þe yere
& v hennes & a cok for iij s. in þe yere þ̵ be som dayis & oþ̵
þᵗ will say nay to þis approument but i shall preve you by
goode resone þᵗ it may be for in halfe a yere be xxvj wekis
& in þis xxvj wekes be ix^xx dayis & in eche off these dayis
ye shall haue one egge off eche henne & þᵗ is ix^xx egge off
oũy henne in þᵗ halfe yere it is a feble soyle of ege &
xxx ege be not worthe vj d. & yeff eny off þem set in þe
halfe yere or pventure som day in defaute off laying ye
shall sufesyently be recompensed þer fore & off vj more for
to bere ovt þe ferme off þe cok withe þe sayle off þe cheknys
þᵗ yo⁼ sytyng henne hathe broughte furthe & withe þe
sayle off þe egis þᵗ yo⁼ hennes shall lay in þᵗ oþ̵ halfe yere
nowe may ye se wheder it may be or no.

The pecok shall answere you off as moche for his fedris
as þe shepe for his woll. Eũy cowe shall answere you off
a calfe and eũy moder shepe shall answere you of a lambe.
Eũy female swynne shall answere you xiij pigis at iij
farowinge & at iij times iiij at eche tyme & at þe iij tyme v
& þe x for tethe. Eũy henne shall answere you off ix^xx
egis or chekyns to þe valewe or better. Eũy goose off vj
goselynge & yeff eny off these catell or foulis be baren þe
bayle shall answere off þe issue þᵗ is loste throughe his evyl
kepynge by cawse he dit not þem & put þe sylũ to oþer
approwment to þe valewe.

The xv chapitur.

Nowe by and sell in seasone & lok þᵗ ye haue trewe
men to hire your bargon so þᵗ ye haue recorde yeff myster

be & take good hede to yo⁼ bayle ffor it is often tymis
seyne þᵗ he encresithe som off soyche þyngs as he byethe &
selithe fro his maystur & p̄ for yeff the bergan be not good
let hym kepe it to hym selfe & answere off þe very vallue
p̄ of bothe in pryse & in approumentę yeff he by & sell by
be weyght & be ware off hym þᵗ holdithe þe balaunce for he
may do you grete fraude it is to wite þat a j d. englyshe
withovt tonchoure oweth to wey xxx graynes off whehete
in þe mydist off þe ere & xx d. englishe owen to wey anowce
& xij ownce is in a pound & viij pound makithe a galon
& viij galons makithe a bushell & viij bushelis makithe a
quarter.

The xvj chapitur.

Nowe off accompte loke þᵗ it be made onys in þe yere
for it was firste ordeyned to wite & to knowe þe value &
state off your maneris & to haue knowlege off all maner off
issues in byinge & sellyng & dispensis as welle off howsold
as off all oþer þyngę & yeff ye haue eny rente or money
loke ye take it ovt off yo⁼ offeceris hondis onys a yere ffor
it is often seyne þᵗ þe bayle & off þe lordis offecers make p̄
marchandise withe p̄ lordis sylū to p̄ owne avayle & grete
hurt to þe lorde & yeff p̄ be eny arrerage loke ye geder þem
withe ovten delay & lok ye take þe namys off þem þᵗ owe þe
rerage for it is often seyne þᵗ þe bayle is seyvere hym selfe
& by þᵗ he makithe oþ̄ reseyvours vnder hym & so one
sparithe another so þᵗ þe arrerage ben often tymis forgeten
& loke þᵗ ye viset your þyngę wisely & often & þen ye shall
see yeff þey take eny harm & amende þen also loke you
visite often tymis yo⁼ servauntę & what shall cause þem to
be more ware off doinge amys & to be more besye abovte
p̄ werke & youris.

Explicit þe tretyce off hosbandrye.

HOSEBONDERIE

AUTHOR UNKNOWN

CEO EST HOSEBONDERIE.

Cest escrit si aprent la manere coment hom deit charger bailifs & prouoz sur lur aconte rendre de un maner E coment hom deit maner garder.

Al primer deit celuy ke rente aconte iurrer ke il rendra leal aconte & leaument se chargera de quant ke il ad receu des biens le seyngnur ne riens ne metra en sun roule fors ceo ke il ad leaument despendu & a prou le seyngnur a sun ascient. E le clerk iorra ke il ad leaument entre en sun roule ceo ke il entent ke sun mestre eit receu des biens le seyngnur ne rien nad entre en sun roule fors ceo ke il entent ke seit a prou le seyngnur. E puis si il ad autrefeez rendu aconte ver coment il pari & si il est troue en arrerages de deners ou de ble ou de estor ou de nule autre chose metre le tot en certeyne value de deners & charger le tot a comencement de sun roule et puis charger le de toutes autres resseites de rentes assise & de toutes autres choses dount nul denor puisse leuer & charger le del tot & mettre le au drein en vne grosse summe & puis aler a despenses.

De coust de charettes.

Primes al coust des charettes bon est ke le feure prenge vn certeyn pur trouer quant ke couent de fer & de acer a charues et ferrer les cheuaus & les affres de leynz a mieuz ke lem poet couenaunt fere a luy sulom ceo ke hom doune ailors en le pays. E si deit hom ver si lia en maner fust cressaunt ou boys gros ou merin gros[1] ke hom puisse

[1] [ou menu.]

THIS IS HUSBANDRY.

This writing teaches the way in which a man ought to direct bailiffs and provosts about rendering the account of a manor, and how a man ought to look after a manor.

In the first place he who renders account ought to swear that he will render a lawful account and faithfully account for what he has received of the goods of his lord, and that he will put nothing in his roll save what he has, to his knowledge, spent lawfully and to his lord's profit. And the clerk shall swear that he has lawfully entered in his roll what he understands his master has received of the lord's goods, and has entered nothing in the roll but what he understands may be to the profit of the lord. And then if he has rendered account before see how it compares, and if he is found in arrears of money, or of corn, or of stock, or of any other thing, put the whole in a stated money valuation, and charge it at the commencement of his roll, and then charge it with all other receipts of assise rents and with all other things for which any money can be raised, and charge it with the whole and put it at the end in a sum total and then go on to expenses.

THE COST OF CARTS.

First as to the cost of carts. It is right that the smith should take a certain sum to find what is necessary in iron and steel for the ploughs and to shoe the horses and avers of the place, as well as one can bargain with him, according to what is paid elsewhere in the country. And then it must be seen if there is underwood on the manor, or large wood

prendre verges et harz & autres choses necessaires saunz
achat ausibien a charettes cum a charues si le prenge lem
pur esparnier les deners e du surplusage ke nest mye troue
ke luy couent achater luy seit alowe par resun e bon
serreit ke il puisse auer charetters & charuers ke seussent
ourir tot lur merin demeyne tot luy deiuent le plus cher
lower. Hom deit ver al chef del an toutes les menues
choses necessaires & touz les estors toutes les ferrures &
toutes les choses ke en le maner demorent petites ou
graundes & mettre en escrit ke hom puisse ver al autre an
queu chose luy couent bosoyngnablement achater & ceo
alower & le surplusage retrere.

Le office du prouost.

Le prouost deit fere quillir tot le peil des affres a fere
cordes a ceo ke il auera a fere. E il deit fere semer en la
cort caumbre a fere cordes a charettes & a cheuestres & as
autres chose bosoyngnables & luy deit estre alowe le fere
si lia nul en la cort ki les sache fere. Amendement de
mesuns de murs de hayes & de fossees si mester seit luy
deit estre alowe sulom ceo ke resun seit. Ne le prouost ne
deit riens vendre ne achater ne resseiure ne liurer si par
taile non & par bone temoyngnaunce. E deit le prouost
fere touz les serianz de la cort quant il venent de lur labor
ourer en la cort de batre ble ou fere murs ou fossees ou
hayes ou autres ourayngnes [1] en la cort pur esparnier le
dener. E si li ad seriaunt ke sache fere ourayngne en la
cort dont luy couendreit alower cher vn autre si li face fere
cel ourayngne & lowe vn autre en sun lu. Les seneschaus
ou les chefs bailifs deiuet ver touz les achaz & toutes les
ventes ke les prouoz ou les suzbailifs funt pur ver ke il
serent bien fet & a prou le seyngnur. E deiuent les
seneschaus & les chefs bailifs ke tenent cort tauntost apres
la seint michel rendre sus lur roules [2] au seyngnur ou al
auditor del aconte ke il puissent charger par ces roules les

[1] [necessairres.]   [2] [de la court.]

or large timber that can be taken for poles and harrows and other necessary things without buying as well for carts as for ploughs, let them be taken to save money, and for the rest which is not found but what he must buy let him be paid justly. It would be well if he could have carters and ploughmen who should know how to work all their own wood, although it should be necessary to pay them more. At the end of the year ought all the necessary small things and stock, horseshoes, and all things which belong to the manor, small or great, to be seen and put in writing, that one may see another year what thing must necessarily be bought, and make allowance for it and subtract the surplus.

The office of provost.

The provost must cause all the hair of the avers to be gathered to make ropes for which he shall have need, and he must cause hemp to be sown in the court to make ropes for the waggons, for harness and other necessary things, and an allowance must be paid for making them if there is any one in the court who knows how to do so. For repairing houses, walls, hedges, and ditches if need be an allowance must be paid according to what is right. And the provost must not buy, sell, receive, or deliver anything unless by tally and good witness. And the provost must make all the servants of the court when they come for their labour work in the court in threshing corn or making walls or ditches or hedges or other works in the court to save money. And if there is a servant who knows how to do work in the court for which it would be necessary to pay another highly, let him do the work and pay another in his place. The seneschals or head-bailiffs ought to see all purchases and all sales that the provosts or under-bailiffs make to see that they are well made and to the lord's profit.

And the seneschals and chief bailiffs who hold court must, immediately after Michaelmas, give up their rolls to the lord or the auditor of the account that they may be able to charge

prouoz & les bailifs ke deyuent aconte de purchaz de cors de tot lan. E le prouost deit respondre del issue des iumentes de la cort ceo est asauer de chescune iumente vn poleyn par an e si li eit nule ke neyt poleyn si seit enquis si ceo seit par mauueyse garde ou par defaute de viaunde ou par trop graunt trauail ou par defaute de stalun ou ke ele seit barayngne ke ele ne porte mye poleyn & la poeit auer chaunge pur vn autre a tens & ne fist mye si seit charge pleynement del issue ou de la value. E si ili eit nul cheual ou beste morte en la cort si seit enquis si ceo seit par defaute de gardeyn ou du bailif ou du prouost ke le puissent auer sauue ou nul amendement auer mys & ne firent si le paent de lur e si il morent par meschaunce ke il ne puissent mes si com morine auient acoune feez sur bestes respoyngne le prouost des quirs & des peaus & des chars & des issues & le metre al emprowement le seyngnur a myeuz ke il sauera ou purra. E si li eit nule chose perdue en la cort ou dehors ou emble lequel ke ele seit viue ou morte petite ou graunde ou le seyngnur puisse auer nule manere de perte ou par arsun ou en nule autre manere le seyngnur deit prendre a prouost & le prouost deit prendre a ceus de la cort ke copables seient. E fet a sauer ke touz les seriaunz de la cort hommes & femmes deiuent estre entendanz a prouost pur ceo ke le prouost deit respoundre de lur feez & le prouost deit mettre ceus en la cort pur queus il voudra respoundre de lur fetz. E deit le seneschal ver ke le prouost eit bons plegges de touz ceus de la cort ki par le prouost isunt mis e si le seyngnur resseit nul damage par le prouost & le prouost neit dont il puisse rendre les damages touz ceus de la vile ke luy elurent rendent le surplus pur luy de ceo ke il ne purra paer. E si le seyngnur ymette parker ou messer ou graunger ou autre ki ke il seit & le seyngnur resseiue nul damage par nul de eus ke il imette le seyngnur se deit prendre a eus pur ceo ke il imist & nyent a prouost. Fet a entendre ky as maners ke sunt gardes par bailifs[1] ke respount del maner tot autresi com le prouost rende

---

[1] [ou ili a nul prouost fors baillif.]

by these rolls the provosts and bailiffs who must account for the purchases of the court throughout the year. And the provost must answer for the issue of the mares of the court, that is to say, for each mare one foal in the year, and if there be any which has no foal let it be inquired if it be by bad keeping, or want of food, or too hard work, or want of stallion, or because it was barren, that she bore no foal; and if she could have been changed for another in time and it was not done, let him be charged fully for the issue or the value. And if there be any horse or beast dead in the court, let it be inquired if it was for want of keeping or because the bailiff and provost could have saved it or made any amendment and did not let them pay it themselves, and if they died by mishap that they could not help, as murrain which falls sometimes on beasts, the provost must answer for the skins and hides and flesh and issues, and put it to the profit of the lord as well as he knows or is able. And if there be anything lost in the court or without, or stolen, whether it be live or dead, small or great, where the lord can have any kind of loss, either by fire or any other way, the lord must take [the value] from the provost and the provost must take it from those of the court who may be to blame. And make it known that all the servants of the court, men and women, ought to obey the provost, because he must answer for their doings, and the provost must put those in the court for whose doings he will be answerable. And the seneschal must see that the provost has good pledges for all those in the court who are put there by him, and if the lord receive any damage by the provost, and the provost cannot make good the damage, all those of the township who elected him shall make up for him the amount he cannot pay. And if the lord place any parker or messer or granger or other, whoever he be, and the lord receive damage from any of these he places, he must take the value from them, because he put them there, and nothing from the provost. Make it known that on the manors which are kept by bailiffs they must answer for the manor, just as the provost renders account even so must

aconte tot autresi deit respondre a toutes choses ne mot remue ne chaunge cum le prouost. Touz ceus ke tenent en villenage de vn maner deiuent elire le prouost tiel pur ke il voleyent respoundre kar si le seyngnur resseiue nul damage par defaute de prouost & il[1] ne eit mie du sien dont il le puisse rendre ke il paent pur luy le surplus ky il ne purra paer.

Respons du semail.

Toute la terre deit estre mesuree en chescun champ par sey & chescune coture du champ nome par sun non e chescun pre par sey & chescune pasture & chescun boys & chescune launde & turberie & more & marreys ausi par sey & tot par la perche de xvi peez & demy pur ceo ke hom poet par resun la terre ke est mesuree par la verge de xvi peez & demy semer en mouz de lius iiii acres de vn quarter & en mouz de lius couent il vn quarter & demy a semer v acres de furment & de segle & de feues & de poys & deus acres de vn quarter de orge & de aueyngne mes pur ceo ke les vnes terres volunt estre semez plus espes ke les autres si deit hom en chescun maner mesurer i acre pur chescun ble & ver de combien hom purra semer chescune manere de ble sur vne acre mesuree & donc poet estre certeyns toz iors de vostre semail. E pur ceo ke hom seme le erge[2] en le chaump de furment & les feues & les poys & les lentiles & les aueyngnes si deit hom nomer chescune coture ke est seme de orge entre le furment & chescune coture de autre ble ke est seme entre les aueyngnes. E la ou les champs sunt parti en ij le iuernage & le trames est tot seme en j champ donc deit il respoundre de chescune coture quele coture est seme de vn ble & quele de autre. E si liad inhom il deit ver quele coture il prent en le inhom & de quel ble il seme chescune coture & tel semail deit il tailer tot par luy & respondre tot par luy hors des autres blez.

---

[1] le prouost.   [2] orge.

he render account for everything, and move and change nothing as the provost. All those who hold in villenage on a manor must elect as provost such a one as they will answer for, for if the lord suffer any loss by the fault of the provost, and he have not of his own goods the wherewithal to make it good, they shall pay for him the surplus which he cannot pay.

THE RETURN FOR SEED SOWN.

All the land ought to be measured in each field by itself and each cultura of the field named by its name, and each meadow by itself, and each pasture and each wood and each waste and turbary and moor and marsh also by themselves, and all by the perch of sixteen feet and a half, because one can in many places reasonably sow four acres with a quarter of seed, where the land is measured by the perch of sixteen feet and a half, and in many places it requires a quarter and a half to sow five acres with wheat, rye, and beans and peas, and two acres with a quarter of barley and oats, but, because some lands must be sown more broadly than others, let there be measured on each manor an acre for each corn, and see with how much one can sow each kind of corn on a measured acre, and thereby can you always be sure of your corn. And because barley is sown in a wheat field and peas and vetches and oats, therefore each cultura which is sown with barley among the wheat must be named and each division of other corn which is sown among the oats. And there where the fields are divided in two, winter seed and spring seed are both sown in one field, for which each division must answer as it was sown with one corn or another. And if there is inhom it must be seen what cultura he takes in inhom, and with what corn he sows each cultura, and such sowing he must tally all by itself and answer for all by itself apart from the other corn.

Coment hóm deit alower les ourors en aust & en tens de fenesun.

Vous purrer bien auer sercle iij acres pur i d. e auer fauche lacre de pre pur iiij d. e lacre de pre de wauz pur iij d. ob. et torner e leuer lacre de pre pur i d. ob. E torner & leuer lacre de wauz pur i d. q. E vos deuet sauer ke v hommes pount bien sier & lier ij acres le ior de chescune manere de ble ke¹ le vn plus & lautre meyns. E la ou chescun prent ij d. le ior si deuet doner pur lacre v d. E la ou les iiij prenent chescun i d. ob. le ior & le quint par ceo ke il est lior ij d. le ior donc deuet doner pur lacre iiij d. E pur ceo ke en mouz de pays il ne seuent nyent sier par lacre si poet hom sauer par les siors & par les iornees ceo ke il funt mes ke vos reteyngnet les siors par les eez. Ceo est asauer ke v hommes ou femmes lequel ke vos voudrez ke hom apelle ii hommes funt vn eez e xxv hommes funt v eez e pount xxv hommes sier & lier x acres le ior enter ouerable e en x iors c acres e en xx iors cc acres par v^{xx} & veet dont quantes acres ilia a sier par tot & veet si il se acordent as iornees & alowet les donc e si il acontent plus des iornees ke ne affert sulom cest aconte si ne lur deuet pas alower kar ceo est en lur defaute ke il ne vnt pas sie les ourayngnes ne il ne vnt pas fet ourer si bien com il deussent.

Coment la terre deit estre mesuree.

E pur ceo ke les acres ne sunt mye touz de vne mesure kar en acon pays mesurent il par la verge de xviii peez e en acon par la verge de xx peez e en acoun par la verge de xxij peez e en acoun par la verge de xxiiij peez. E deuet sauer ke lacre ke est mesuree par la verge de xviii peez fet i acre & vne rode & le xvj de vne rode de la verge de la verge² de xvi peez e iiij acres funt v acres & le quart de vne

---
¹ de.   ² Sic.

How one must pay labourers in August and in time of haymaking.

You can well have three acres weeded for a penny, and an acre of meadow mown for fourpence, and an acre of waste meadow for threepence-halfpenny, and an acre of meadow turned and raised for a penny-halfpenny, and an acre of waste for a penny-farthing. And know that five men can well reap and bind two acres a day of each kind of corn, more or less. And where each takes twopence a day then you must give fivepence an acre, and when four take a penny-halfpenny a day and the fifth twopence, because he is binder, then you must give fourpence for the acre. And, because in many places they do not reap by the acre, one can know by the reapers and by the work done what they do, but keep the reapers by the band, that is to say, that five men or women, whichever you will, who are called half men, make a band, and twenty-five men make five bands, and twenty-five men can reap and bind ten acres a day working all day, and in ten days a hundred acres, and in twenty days two hundred acres by five score. And see then how many acres there are to reap throughout, and see if they agree with the days and pay them then, and if they account for more days than is right according to this reckoning, do not let them be paid, for it is their fault that they have not reaped the amount and have not worked so well as they ought.

How the land ought to be measured.

Because acres are not all of one measure, for in some countries they measure by the perch of eighteen feet, and in some by the perch of twenty feet, and in some by the perch of twenty-two feet, and in some by the perch of twenty-four feet, know that the acre which is measured by the perch of eighteen feet makes an acre and a rood, and the sixteenth of a rood, of the perch of sixteen feet, and four acres make five acres and a quarter of a rood, and

rode e viij acres funt x acres & demye rode. E xvj acres funt xx acres & une rode. E lacre ke est mesuree par la verge de xx piez fet i acre & demye & le quart de vne rode e iiij acres funt vi acres & une rode e viij acres funt xii acres & demye. E xvi acres funt xxv acres. E lacre ke est mesuree par la verge de xxij peez fet i acre & demi & une rode & demi & le xvi de vne rode e les iiij acres funt vii acres & di. & le quart de vne rode. E les viij acres funt xv acres & demye rode. E les xvi acres funt xxx acres & une rode. E lacre ke est mesuree par la verge de xxiiij peez fet ij acres & une rode. E iiij acres funt ix acres.

Respons del issue de la graunge.

Del issue de la graunge deit hom ver combien il iad seme de chescun ble & de combien il respount del issue ke par dreit & par comoune respounse le orge deit respondre al viii greyn ceo est asauer de vn quarter seme viij quarters del issue. E le segle al vij greyn e feues au vi & poys a vi. E de drage de orge & de aueyngne si il est ouelement medle au vi. E si lia plus de orge ke de aueyngne plus deit respoundre. E silia meyns de orge ke de aueyngne le moyns. E ausi de mestilon de furment & de segle si il est ouelement medle deit respoundre au sime e si lia plus de segle ke de furment le plus deit respoundre. E silia plus de furment ke de segle le meyns. E le furment par dreit deit respondre a v greyn & le aueyne au quart mes pur ceo ke les terres ne respounent ausibien vn an cum vn autre ne les mauueyses terres ne respounent mye cum les bones. E de autre part il auent ke le iuernage se prent bien & le trames faut e acoune feez le trames se prent bien & le iuernage faut. E pur ceo si la terre ne respount de plus ke ele nest charge par le greyn le seyngnur iperd. E si ele respoyngne de meyns il couent ke celuy ke rent laconte le paye del sien demeyne. E pur ceo ne poet hom mye prendre certeynement a la respounse auaunt dite e ne mye pur ceo mouz de genz le pernent issin ke par le

eight acres make ten acres and a half rood, and sixteen acres make twenty acres and a rood. And the acre which is measured by the perch of twenty feet makes one acre and a half and the quarter of a rood, and four acres make six acres and a rood, and eight acres make twelve acres and a half, and sixteen acres are twenty-five acres. And the acre which is measured by the perch of twenty-two feet makes one acre and a half, and a rood and a half and the sixteenth of a rood, and four acres make seven and a half and quarter of a rood, and eight acres make fifteen acres and a half rood, and sixteen acres make thirty acres and a rood. And the acre which is measured by the perch of twenty-four feet makes two acres and a rood, and four acres make nine acres.

THE RETURN FROM THE PRODUCTS OF THE GRANGE.

As to the issue of the grange, one must see how much there is sown of each corn and how much it yields for issue by right and common return; barley ought to yield to the eighth grain, that is to say, a quarter sown should yield eight quarters; rye should yield to the seventh grain, and beans and peas to the sixth. And dredge of barley and oats, if equally mixed, to the sixth, but if there is more barley than oats it ought to yield more, and if there is less barley than oats, less. And also the mixtelyn of wheat and rye, if it is equally mixed it should yield to the sixth, and if there is more rye than wheat it ought to yield more, and if there is more wheat than rye, less. And wheat ought by right to yield to the fifth grain and oats to the fourth, but because lands do not yield so well one year as another, nor poor land as the good, and besides it may happen that the winter sowing takes well and the spring sowing fails, and sometimes the spring sowing takes well and the winter sowing fails, and because, if the land does not yield more than was sown, then the lord loses, and if it yield less he who renders account pays it himself. And so one cannot be sure of the yield above mentioned and not because many people take

greyn. E ki ne veut issi si metre vn leal homme
en ki il se affie outre la baterie de la graunge. E bon est
ke celuy ke est outre la baterie metre en la taile le issue
de chescune meye & de la graunge par sei pur ver de quants
de quarters chescune meye respount par sei. E si li ad
tas de hors si le face mesurer par rode & par pee la leaure
& la lungure & la hautor quant il le fra batre & taile
chescun tas par sei. E donc purra il sauer ausibien de
chescune meye de la graunge com de chescun tas dehors
la response & le issue mes ke les tas seient chescun an de
vne mesure de leaure & de lungure & de hautor. E si il
veut vendre sun ble en gres si poet il le myeuz sauer combien chescun tas deit valer sulom le marche du ble. E tot
vende il le ble en gres bon est ke il taile & ke il veie le
issue de chescune meye & de chescun tas kar com plus
souent esprouera plus certeyn serra del issue & de la respounse pur ceo ke les blez ne respounent pas chescun
an ouelement. E prenge garde celuy ke est outre la
baterie de ble ke si il bate nul viel ble entre le nouel ke
il batre & taile le viel tot par luy e ke le prouost respoyngne en sun roule del vente de ble tot par luy pur
ver le issue de chescune annee si il respount a sun semail.
E si vos fetes brez il vos deit toziors respondre de ix
quarters le x[1] a tot le meyns & si est ceo mout petite
response mes hom le met a ceo pur ceo ke hom purra
fere le ble trop germir pur fere vn graunt respouns del
auantage par vnt le brez vaudreit mout le meyns & le
meyns respoundreit de ceruoyse.

Response de la daerie e coment la dae deit respondre del menu estor de la cort & de lur issue.

E vos deuet auer en chescun lu ou daerie est homme
ou femme pur garder le menu estor de leinz cum auant
est dit. E si il est homme si deit il fere toutes choses
cum vne femme ifust e deit prendre a xvi semeyngnes le

[1] dime.

it so by the grain. And he who does not wish it so, let him put a true man in whom he trusts over the threshing. And it is well that he who is over the threshing should tally the product of each mow of the grange by itself to see how many quarters each mow yields by itself. And if there be a stack outside, let it be measured by rod and by foot, the breadth, length, and height, when it is about to be threshed, and tally each stack by itself, and then it will be possible to know the yield and issue of each mow outside as well as of each stack within the grange; but let the stacks be each year of the same size in breadth and length and height. And if he wish to sell his corn in gross, he will know better how much each stack is worth according to the price of corn. And although he sell the corn in gross, it is well to tally it and see the issue of each mow and of each stack, as the more often he proves it the more sure he will be of the yield, because corn does not yield equally each year. Take care that he who is over the threshing, if he thresh any old corn among the new, that he thresh and tally the old quite by itself; and let the provost answer in his roll for the sale of the corn quite by itself, to see the issue of each year, if it yields its seed. And if you make malt, he must always answer you, for nine quarters, a tenth at the least; and this is a little yield, but it is fixed thus because corn can be made to sprout too much to make a good return for profit, whereby the malt is worth much less and will yield less ale.

The yield from the dairy, and how the dairywoman ought to answer for the small live stock of the court and for their issue.

You must have, in each place where there is a dairy, a man or woman to keep the small live stock there, as said before. If it is a man, he must do everything as a woman would, and he ought to take every sixteen weeks a quarter

quarter pur le auantage ke il ad du blaunk la ou les autres serianz prenent a xii semeyngnes. E ele deit venter tot le ble & serra de la moyte del ior pae pur paer la femme ke la aide. E il deiuent venter quatre quarters de furment ou de segle e vi quarters de orge & de poys & de feues & de oriace pur i d. E viii quarters de aueyngne pur i d. E hom deit toziors prendre au quart le v outre pur le comble de toute manere de ble. Ausi deit hom batre le quarter de furment ou de segle pur ii d. e le quarter de orge & de poys & de feues pur i d. ob. & le quarter de aueyngne pur i d. & toziors alower au quart le quint pur le comble. E si deit la dae prendre garde a tot le petit estor ke demoert en la cort cum de purceaus letanz & de pouns & de lur issue.e de owes & de lur issue [1] & de chapons & de coks & de gelines & de poucins & des oefs & de lur issue. E vos deuet sauer ke la troie deit purceler par ii fez en lan a chescune feez au meyns vii purceaus. E chescune owe v oisons par an. E chescune geline de cxv oefs vii poućins dont les iij deiuent estre fet chapons. E si li ad trop de poucins femeles si les chaunge pur males tant cum il sunt iouenes si ke chescune geline puisse respondre de iij chapons & de iiii gelines par an. E couent a v owes i garok & a v gelines i cok. E chescune vache deit respoundre de vn veal par an. E chescune mere berbit de i anignel par an. E si li eit vache ou mere berbit ke neit porte lan si fet a enquerre par ki defaute ceo est ou en le bailif ou en le proust ou en le gardein par defaute de garde ou par defaute de viande en leste ou en le iuer ou par defaute de male ou si le prouost le peust auer change pur autre a tens & nel fist mye. E si il seit troue en nule defaute de eus si seit charge tot pleynement del issue ou de la value. E ausi si nul moert en nule manere par lur defaute si respoyngnent de la beste viue ou de la value. E si ceo est maner ou daerie ne soit mye si est toteueirs bon de auer vne femme leinz a plus leger coust ke hom poet pur prendre garde del menu estor de leinz & de quant ke est dedenz la cort, &

[1] [osyons.]

[of corn], because of the advantage he has from the milk, where other servants take it every twelve weeks. And she must winnow all the corn, and shall be paid for a half-day to pay the woman who helps her. And she ought to winnow four quarters of wheat or of rye and six quarters of barley and peas and beans and oriace for a penny, and eight quarters of oats for a penny. And one must always take for four a fifth over for the comble of all kind of corn. Also, one ought to thresh a quarter of wheat or rye for twopence, and a quarter of barley, and peas, and beans, for a penny-halfpenny, and a quarter of oats for a penny, and always allow for four a fifth for the comble. And the dairywoman must take care of all the small animals in the court, as sucking-pigs and peacocks and their issue, and geese and their issue, and capons and cocks and hens and chickens and eggs and their issue. And you must know that a sow ought to farrow twice a-year, having each time at the least seven pigs, and each goose five goslings a-year; and each hen, for a hundred and fifteen eggs, seven chickens, three of which ought to be made capons, and, if there be too many hen chickens, let them be changed for cocks while they are young, so that each hen may answer for three capons and four hens a-year. And for five geese you must have one gander, and for five hens one cock. And each cow ought to answer for a calf a-year, and each ewe one lamb a-year; and if there be a cow which has not calved or a ewe which has not lambed in the year, let it be inquired whose fault this is, either the bailiff's or the provost's or the keeper's, for want of keeping or want of food in the summer or winter, or want of a male, or if the provost could have changed it for another in time and did not, and, if it be found to be any fault of these, let them be fully charged for the issue or its value. And also if any [beast] die in any way by their fault, let them answer for the live beast or its value. And if this is a manor where there is no dairy, it is always good to have a woman there, at a much less cost than a man, to keep the small animals there and what there is within the court, and answer for all pro-

respondre de toutes les issues de leinz cum dae ceo est a
sauer cum de troyes purcelez & de lur purceaus. E de
pouns & de lur poucins si lia. E de owes & de lur osions
de chapons de coks de gelines & de lur poucins & de lur
oefs & deit respoundre de la moite de venter du ble ausi
com la dae.

RESPONSE DE VACHES DE GENICES & DE LUR BLAUNK.

Chescune vache deit respoundre de len demeyn de la
seint michel iekes les primeres kalendes de may par xxviii
semeyngnes le vn ior & lautre toute de x d. pur tot cel tens
ke le vn plus & lautre meyns. E fet a entendre ke toutes
les vaches ne respounent pas ouelement, les vnes responent
plus & les autres meyns & les vnes sunt plustost leteres ke
les autres & plustost sekes. Ne les genices ne rendent pas
ataunt de leat a lur primere portur cum il funt al autre
porture apres mes lune parmy lautre deit atant respondre
par resun. E de len demeyn de les primeres kalendes de
may iekes le ior seint michel par xxiiij semeyngnes le vn
ior & lautre conte & funt viij^{xx} & viii iors & deit valer le
issue de leat de chescune vache par cel tens iij s. vj d. del
issue de chescune vache. E toute lautre sesun amoute le
issue de la vache a x d. E par cel aconte deit chescune
vache respondre de iiij s. iiij d. del issue de leat. E fet a
sauer ke chescune vache deit respondre entre les kalendes
de may & la seint michel de vi peres de furmage & de
atant de bure cum affert ataunt de furmage ceo est a
sauer toziors a vij peres de furmage j pere de bure. E le
deit fere toziors furmage de len demyn de seynt michel
iekes le ior seint martin au meyns mes en lautre sesun
apres nouel iekes le este si il plus profite au seyngnur de
vendre leat ke de fere furmage kar ataunt vaut a vendre i
galon de leat a donc cum iij en este ou en vne autre sesun.
E si vos feisses furmage donc ne vaudreit i galon de leat
adonc nyent plus ke en autre sesun.

duce there as a dairywoman would—that is to say, when the sows farrow for their pigs, for peacocks and their chicks, if there are any, for geese and their goslings, for capons, cocks, hens, and their chickens and eggs—and she ought to answer for the half of the winnowing of the corn also as the dairywoman.

The return from cows, heifers, and their milk.

Each cow ought to yield, from the day after Michaelmas until the first kalends of May, for twenty-eight weeks, one day with another, tenpence for all that time, more or less. And it must be understood that all cows do not yield alike; some give more, some less, some give milk sooner than others and are sooner dry, and heifers do not give as much milk at their first bearing as after, but, one with another, they ought to yield as much reasonably. And from the morrow after the first kalends of May until Michaelmas, for twenty-four weeks, one day counted with another, makes one hundred and sixty-eight days; then ought the yield of milk of each cow to be worth, during that time, three shillings and sixpence, and all the other season the yield is worth tenpence, and by this reckoning each cow ought to yield milk to the value of four shillings and fourpence. And be it known that each cow should give as well, from the kalends of May to Michaelmas, six stones of cheese and as much butter as shall make as much cheese—that is to say, always, to seven stones of cheese, one stone of butter. And cheese should always be made from the morrow after Michaelmas until Martinmas, at least; but in the other season, after Christmas until the summer, it is more profitable to the lord to sell the milk than to make cheese, for then a gallon of milk, if sold, is worth as much as three in summer or at another time. And if you should make cheese, then a gallon of milk is not worth more than at another time.

RESPONS DE BERBIZ & DE LUR BLAUNK.

Chescune mere berbit deit respondre del issue de sun leat partot le este tant cum ele est letere de vi d. Kar les meres berbiz ne sunt mye leteres outre le aust, ne hom ne les tient mie volunters leteres[1] aust pur ceo ke il valent le meyns & sunt plus perilouses a iuerner, e si eles seient malades ou febles si letent le meyns. E si deit la dae respoundre de autretaunt del issue de vn galon de leat de furmage & de bure de berbiz cum de i galon et demy de leat de vache. E i galon de bure peyse vij li. e ii galons peisent xiiii li. e xiiij li. funt la pere and xiiij peres funt la wae. E fet a sauer ke la iumente vet xlix semeynes dekes ele eit polenc. E la vache vet xl semeynes dekes ele eit velee. E la berbit del houre ke ele est assaille dekes ele eit aingnele xxi semeyngnes. E la treie vet del houre ke ele est assaille xvj iekes ele eit purcele. E la troi poet purceler v feez en ij aunz & nyent plus. E le owe coue vne feez par an si ele est bone mes ceo ne fra ele mie chescun an ne chescune ne poet mie fere issi, mes sulom ceo ke eles seient bien gardez si respondrent de plus ou de meyns.

COMENT HOM DEIT METTRE LE ISSUE DE SUN ESTOR A FERME.

E si vos volet mettre le issue de vostre estor a fermes vos deuet prendre de chescune vache iij s. & vj d. de cler & aquiter la dime & sauuer vos la vache & le veal. E de la berbit vj d. & aquitter la dime & sauuer vos la berbit & le aignel. E la troie vos deit rendre vj s. vj d. par an & aquiter la dime & sauuer vos la troie. E chescune mere owe vos deit rendre vij d. ob. de cler & aquiter la dime & sauuer le owe. E chescon geline vos deit rendre par [[2]] de cler & aquiter la dime & sauuer la geline. E x quarters mmes][3] & de peres deiuet respoundre de vii tonel de

[1] [apres.]   [2] illegible [an ix d.]   [3] de poumes.

THE RETURN FROM SHEEP AND THEIR MILK.

Each ewe should answer sixpence for the yield of its milk through the summer, while it is giving milk, for ewes do not give milk after August; and no one would willingly have them give milk after August, because they would be worth less and more difficult to keep in winter. And if they be sick or weak, let them be milked less. And the dairywoman ought to answer besides for the yield of a gallon of milk, cheese, and butter from the sheep, as a gallon and a half of milk from the cow. And a gallon of butter weighs seven pounds, and two gallons weigh fourteen pounds, and fourteen pounds are a stone, and fourteen stone are a wey. And let it be known that a mare is in foal forty-nine weeks, and a cow is in calf forty weeks; a ewe goes with lamb twenty-one weeks, and a sow can farrow five times in two years and not more; and a goose will hatch once a-year if she is good, but she will not do this every year, nor can she be made to, but, according as they are well kept, they will yield more or less.

HOW ONE OUGHT TO FARM OUT THE ISSUE OF THE STOCK.

If you wish to farm out the issue of your stock, you can take four-and-sixpence clear for each cow and acquit the tithe, and save for yourself the cow and calf; and for a sheep sixpence and acquit the tithe, and keep the sheep and lamb; and a sow should bring you six shillings and sixpence a-year and acquit the tithe, and save for yourself the sow; and each goose ought to bring you sevenpence-halfpenny clear and acquit the tithe and save the goose; and each hen should bring you ninepence clear and acquit the tithe and save the hen. And ten quarters of apples and pears should yield seven tuns of cider; and a quarter

cisere. E i quarter de neiz deit respondre de iiij galons de oile. E chescune rouche de eez deit respoundre de ij rouches par an de lur issue lun parmy lautre kar acoune ne rent nule & acoune iij or iiij par an. E en acon lu lur doune lom a manger rien de tot le iuer e en acon lu lur doune lom. E la ou hom lur doune a manger si pount il pestre viij rouches tot le iuer de i galon de miel par an. E si vos nel quillez fors en ij aunz si aueret ij galons de miel de chescune rouche.

EXPLICIT HOSEBONDERIE.

of nuts should yield four gallons of oil. And each hive of bees ought to yield for two hives a-year, one with another, for some yield nothing and others three or four a-year, and in some places they are given nothing to eat all winter and in some they are fed then, and where they are fed you can feed eight hives all winter with a gallon of honey; and if you only collect the honey every two years, you should have two gallons of honey from each hive.

THE END OF HUSBANDRY.

SENESCHAUCIE

AUTHOR UNKNOWN

Ci comence la seneschaucie ke pertint a seneschal de terres.

Le seneschal de terres deit estre sages e leaus e apruant e deit sauer [lassise del regne][1] pur soreyne bosoignes defendre e pur les baillifs ke desoz li sont en lor dotances certifier e aprendre. Item il deit deus o iij foiz fere son tour par an e visiter les maners de sa baillie e donke deit il enquerre de rentes de seruises e de costumes concelees e sustretes e de franchise de corz de terres e de boys de prez de pastures de ewe de molins e des autre choses ke a maners aportenent sanz garant aloynnez par li[2] e coment e si il ad poer de amender les choses auant dites en forme de dreiture sanz tort fere a nuly[3] [4] e si il ce ne put fere mostre a son seignur ke il se entremette si il neut sa dreiture conquere.

Le seneschal deit a sa premire venue as maners fere mesurer trestoz lor demeynes de chescoin maner par leale genz e il deit sauer par la perche del pays quantes acres il iad en chescun champ e par tant put il sauer combin de forment do seggle de orge de auene de poys de feues de drage len doit par reson semer en chun acre e partant put len uer[5] si le prouost o le hayward acountent plus en semence ke le dreit e partant put il ver quant de charues couent al maner car chescune charue deit par reson arer[6] ix vinz acres ce est a sauer lx al yuernail lx al trameys lx al waret esi put il ver quant des acres deiuent estre are par an de priere e de costume[7] e quant des acres remenent a gaigner des charues del maner e

[1] [les leys.]
[2] ky.
[3] autres.
[4] [il le facet.]
[5] [e sauer.]
[6] [par an.]
[7] [e kans de acres pur denyrs.]

HERE BEGINS THE BOOK OF THE OFFICE OF SENESCHAL.

The seneschal of lands ought to be prudent and faithful and profitable, and he ought to know the law of the realm, to protect his lord's business and to instruct and give assurance to the bailiffs who are beneath him in their difficulties. He ought two or three times a year to make his rounds and visit the manors of his stewardship, and then he ought to inquire about the rents, services, and customs, hidden or withdrawn, and about franchises of courts, lands, woods, meadows, pastures, waters, mills, and other things which belong to the manor and are done away with without warrant, by whom, and how: and if he be able let him amend these things in the right way without doing wrong to any, and if he be not, let him show it to his lord, that he may deal with it if he wish to maintain his right.

The seneschal ought, at his first coming to the manors, to cause all the demesne lands of each to be measured by true men, and he ought to know by the perch of the country how many acres there are in each field, and thereby he can know how much wheat, rye, barley, oats, peas, beans, and dredge one ought by right to sow in each acre, and thereby can one see if the provost or the hayward account for more seed than is right, and thereby can he see how many ploughs are required on the manor, for each plough ought by right to plough nine score acres, that is to say: sixty for winter seed, sixty for spring seed, and sixty in fallow. Also he can see how many acres ought to be ploughed yearly by boon or custom, and how many acres remain to be tilled by the ploughs of the manor. And further he can

partant put il uer quant des acres deiuent estre siez de priere e de costume e quant des acres pur deners e si il iad nul treget en la semence o en le arure o en le sier legirement le aperceura. Item il deit fere mesurer toz les prez e tote les pastures seuerales par acres e partant put hom sauer les costages[1] e combin defein couint par an a la sustenance del maner e de combin de estor lem put sustenir de la pasture seurale e combin de la commune.[2]

Le seneschal ne ad nul poer de remuer baillif ne seriant ke est oue son seignur en chef e a ses robes e a sa liueree sanz especial mandement le seignur car issi freit il de la teste[3] keve[4] mes si le baillif est meins sachant ou meins aprouant ke estre ne deit ou si il ad trespas fet ou mauueste en sa baillie soit mustre a son seignur e a son consail e il enface ce ke il quidra ke bon seit.

Le seneschal ne eit pas poer de garde de mariage ne de eschete vendre[5] ne dame ne femme douer ne homage ne seute[6] prendre ne vilayn[7] vendre ne enfranchir[8] sanz especial garant de son seignur. Item le seneschal ne deit pas estre sourein acontur de choses de[9] sa baillie car il deit sur le aconte de chescon maner respondre de ses fez e de ses comandemenz e de ses apruemenz e de fins e des amerciemenz de curz par ly plede si com vn autre e pur ce ke nul homme ne put ne ne deit estre iuge ne iustice de son fet propre.

Le seneschal deit a sa venue a chescun maner ver e enquere de terres coment ele sont gaignees e ensesonees eles chiuaus charetters e les auers e les boefs e les vaches les berbiz e les porz coment il sont gardes e aprueys e si il yeit perte ou damage par defaute de garge[10] e si deit prendre les amendes de ceus ke sont encupe[11] ensi ke le seignur ne perde. Item le seneschal deit puruer ke chescun maner

[1] [e il deyt sauer.]
[2] [e quel enprouwement en est fet.]
[3] chef.
[4] couwe.
[5] [ne doner.]
[6] fente.
[7] [malle ne femele.]
[8] alegger.
[9] ky touche.
[10] garde.
[11] coupables.

see how many acres ought to be reaped by boon and custom, and how many for money. And if there be any cheating in the sowing, or ploughing, or reaping, he shall easily see it. And he must cause all the meadows and several pastures to be measured by acres, and thereby can one know the cost, and how much hay is necessary every year for the sustenance of the manor, and how much stock can be kept on the several pasture, and how much on the common.

The seneschal has no power to remove a bailiff or servant who is with the lord, and clothed and kept by him, without the special order of the lord, for so he would make of the head the tail; but if the bailiff be less capable or less profitable than he ought to be, or if he have committed trespass or offence in his office, let it be shown to the lord and to his council, and he shall do as he shall think good.

The seneschal should not have power to sell wardship, or marriage, or escheat, nor to dower any lady or woman, nor to take homage or suit, nor to sell or make free a vilein without special warrant from his lord. And the seneschal ought not to be chief accountant for the things of his office, for he ought on the account of each manor to answer for his doings and commands and improvements, and for fines and amerciaments of the courts where he has held pleas as another, because no man can or ought to be judge or justice of his own doings.

The seneschal ought, on his coming to each manor, to see and inquire how they are tilled, and in what crops they are, and how the cart-horses and avers, oxen, cows, sheep, and swine are kept and improved. And if there be loss or damage from want of guard, he ought to take fines from those who are to blame, so that the lord may not lose. The seneschal ought to see that each manor is properly

seit estore a son dreit e si il iad surcarke¹ a nul maner plus ke la pasture ne put soffrir seit remue le surkarke iekes a autres maners la ou id iad meyns de carke e si le seignur ad a fere de deners pur dette rendre o pur achat fere a certein terme auant le terme e auant le tens ke bosoin sorde si deit le seneschal puruer as maners dont il put auer deners a greindre preu e a mendre damage car [purueer ne se uoet si pert souent].²

Le seneschal deit a sa uenue as maners enquere coment le baillif se porte de denz e dehors quele garde il prent e quel apruement il fet e quel encres e quel profit en la maner en sa baillie par la reson de sa demore. E ausins del prouost e del hayward e del estorror e de tuz les autres mesters coment chescun se porte endreit de sey e par³ put il meuz estre certifie ky fet pru e ky damage. Item il deit puruer ke il nyeit a nul maner wast ne destruccion ne nul surcharke de nule chose ke touche le maner il deit remuer toz ceus⁴ ke le seignur ne sont bosoignables e toz les serianz ke de rin ne seruent e tot le surcarke de la dayerie e des autres mesters⁵ sanz preu e reson ke sont a pele fause mises sanz pru.

Le seneschal deit a ses venues a maners enquere de les mesfesanz e de trespas fet en park en viuers en warennes en coningers en columbers⁶ e de toz autres choses ke sont fez en damage de son seignur en sa baillie.

Le office de baillif.

Le baillif deit estre leaus e pruant e bon gaignur e sage ausi ke il ne coueigne mie mander a son seignur ne a son sourein seneschal pur auer consail e aprise de tote les

---

¹ de amaylle.
² [ky auant mayn ne se purueyt souent pert.]
³ tant.
⁴ les serjanz.

⁵ [qe nul prouue font for wast. E il deyt abregger touz les coustages nynt bossognables par les maners.]
⁶ ewes defendues.

stocked, and if there be overcharge on any manor more than the pasture can bear, let the overcharge be moved to another manor where there is less stock. And if the lord be in want of money to pay debts due, or to make a purchase at a particular term, the seneschal ought before the term, and before the time that need arises, to look to the manors from which he can have money at the greatest advantage and smallest loss, for if he will not provide, he will often lose.

The seneschal ought, on his coming to the manors, to inquire how the bailiff bears himself within and without, what care he takes, what improvement he makes, and what increase and profit there is in the manor in his office, because of his being there. And also of the provost, and hayward, and keeper of cattle, and all other offices, how each bears himself towards him, and thereby he can be more sure who makes profit and who harm. Also he ought to provide that there should be no waste or destruction on any manor, or overcharge of anything belonging to the manor. He ought to remove all those that are not necessary for the lord, and all the servants who do nothing, and all overcharge in the dairy, and other profitless and unreasonable offices which are called wrong outlays, without profit.

The seneschal ought, on his coming to the manors, to inquire about wrong-doings and trespasses done in parks, ponds, warrens, conygarths, and dove-houses, and of all other things which are done to the loss of the lord in his office.

The office of bailiff.

The bailiff ought to be faithful and profitable, and a good husbandman, and also prudent, that he need not send to his lord or superior seneschal to have advice and

choses ke touchent sa baillie si ne fust de estrange cas o de grant peril car baillif poy vaut en bosoign ke poi oit[1] e ren nad de sey sanz autrui aprise le baillif deit par matin leuer e soruer les boys les blez les prez e les pastures e ver le damage ke soit fet e il deit uer ke les charues soyent matin ionz e a dreit vre soient desionz[2] ensi ke il facent le iour lor dreit arrure quant ke il fere poent e deyuent par la perche mesuree e il deit fere les terres marler fauder composter aprouer e amender ensi ke son sen apruge pur le pruement e le mendement del maner. Item il deit uer quant des acres mesurees les priez e les custumez deyuent arer par an e quant des acres les charues deiuent gaigner del maner e par tant put il abroger le surplus del cust. E il deit uer e estendre quant des acres de pre[3] les custumes deiuent faucher e leuer e quant des acres de ble les priez e les custumers deyuent sier e carier e par tant put il uer quant des acres de pre remenent a faucher e quant des acres de ble remenent a sier por dener e ensi ne sera il point de ceu de fause alouance e il deit defendre ke nul prouost ne nul bedel ne nul hayward ne nul autre seriant del maner ne chyuauchent ne prestent ne surmeynent les chiuaus charetters ne les autres.[4] Item il deit ver ke les chiuaus e les befs e tot le estor seit bin garde e ke nul autre auers ne pessent ne mangeuent lur pasture.

 Le baillif deit estre dreiturel en toz poinz e en toz se fez e il ne deit pas prendre fin de terre ne de relif ne femme[5] alegger sanz le seneschal ne nule rin pleder[6] fye ne franc tenement[7] ne franchise ke torne a desheriteson de son seignur sanz garant e il ne deit pas remuer prouost ne fere sanz le seneschal mes si il ad fet trespas ou mauueste mettre le par bon plegges li e ses chateus a respondre de ses fez par deuant le seneschal. Item il ne deit a nule manire fere fornir ne batre[8] sanz garant de

[1] seyt.
[2] [de lur ouerayn.]
[3] [les pryeres e.]
[4] affres.
[5] [douwer ne vylyn.]
[6] ky.
[7] [touche.]
[8] brace.

instruction about everything connected with his baillie, unless it be an extraordinary matter, or of great danger; for a bailiff is worth little in time of need who knows nothing, and has nothing in himself without the instruction of another. The bailiff ought to rise every morning and survey the woods, corn, meadows, and pastures, and see what damage may have been done. And he ought to see that the ploughs are yoked in the morning, and unyoked at the right time, so that they may do their proper ploughing every day, as much as they can and ought to do by the measured perch. And he must cause the land to be marled, folded, manured, improved, and amended as his knowledge may approve, for the good and bettering of the manor. He ought to see how many measured acres the boon-tenants and customary-tenants ought to plough yearly, and how many the ploughs of the manor ought to till, and so he may lessen the surplus of the cost. And he ought to see and know how many acres of meadow the customary-tenants ought to mow and make, and how many acres of corn the boon-tenants and customary-tenants ought to reap and carry, and thereby he can see how many acres of meadow remain to be mowed, and how many acres of corn remain to be reaped for money, so that nothing shall be wrongfully paid for. And he ought to forbid any provost or bedel or hayward or any other servant of the manor to ride on, or lend, or ill-treat the cart-horses or others. And he ought to see that the horses and oxen and all the stock are well kept, and that no other animals graze in, or eat their pasture.

The bailiff ought to be just in all points and in all his doings, and he ought not, without warrant, to take fines or relief from the land, nor enfranchise a woman without the seneschal, nor hold pleas touching fees or freehold or franchise which turn to the loss of the lord. And he must not remove or make a provost without the seneschal; but if he have trespassed or done wrong, let him be put in good surety, he and his goods, to answer for his doings before the seneschal. He must not in any wise bake or brew

seignur ne nul ke veigne as maners par le seignur ne sanz le seignur¹ nen eit costages del maner for ce ke le baillif vodra payer de sa burse e si le baillif seit atorne² pur³ deners pur ses estouers ensi ke rin ne prenge del maner for litire fen e busche.

Le baillif deit ver ke il yeit bone garde a granges outre les baturs e ke les blez soient bin e nettement batuz e ke le forage seit bin sauue en bon tas ou en moillong ben couerz e ke nul forage⁴ del maner seit vendu mes soit la litire e feugire si point ya gette en wasseus ou en chimins pur compost fere e nul estuble ne soit uendu a maners mes soit quille ensemble tant com len aura mester a mesons courir e le remeignant demorge desus la terre e seit are oue les charues.

Item le baillif deit suruer les charues les gaignages e surver ke les terres soient bin arez de menu reons e bin ensesonez bin semez de bone e nette semence e bin e nettement hercez e tote la semence yuernage scit achate par garant del bref le seignur ou le seneschal car ce est vn chapitre ke veut auer garant e tot le tramail seit seme del propre si bon marche ne le desturbe par mandement del bref.

Nule choses de maners ke deyuent estre uendues ne seient prises par gent mes soient envoyes as feires e as marchez de plusor lius veue e bargaynnees e ky plus vodra doner si les eyt car ce nest pas chateil de mort ne de guerre ne vente de pin faude le roy.

Nul seneschal ne nul baillif ne nul seriant ne nul prouost ne nul bedel ne nul hayward prenge par pris ne par nule uente des⁵ maners dont eus memes sont gardains⁶ car il ne deiuent pas par reson achater les choses ne prendre par pris ke eus memes deiuent apruer e vendre. Nul baillif ne sofre en sa baillie ke chiual ne auer bof ne vache [jueuene auer]⁷ moton ne mereberbiz ne hogastre soient escorces auant ke lem les eyt veu par quel

¹ sans garant.
² a fer.
³ en.
⁴ ne lytere.
⁵ [choses des.]
⁶ [charge.]
⁷ [jumente.]

without the lord's warrant. And no one who comes to the manor, for the lord or without the lord, may be at the expense of the manor, unless the bailiff wish to pay it from his own purse. And let the bailiff be appointed money wages for his needs, so that he may take nothing from the manor but straw, hay, and firewood.

The bailiff must see that there be good watch at the granges over the threshers, and that the corn be well and cleanly threshed, and that the straw be well saved in good stacks or cocks well covered, and that no forage be sold from the manor, but let the forage and fern, if there be any, be thrown in marshy ground or in roads to make manure. And no stubble should be sold from the manor, but let as much as shall be wanted for thatching be gathered together, and the rest remain on the ground and be ploughed with the ploughs.

And the bailiff ought to oversee the ploughs and the tillage, and see that the lands are well ploughed with small furrows, and properly cropped, and well sown with good and pure seed, and cleanly harrowed; and all the winter seed may be bought by warrant of the writ of the lord or seneschal, for this is a point that must have warrant; and all the spring seed may be sown from his own store, if cheapness does not prevent him by the order from a writ.

Let nothing on the manors which ought to be sold be taken by the people, but let it be sent to fairs and markets at several places, and be inspected and bargained for, and whoever will give the most shall have it; for it is not chattel of death, or of war, or sold from the king's pinfold.

No seneschal or bailiff, or servant, or provost, or bedel, or hayward, should take for money, or through any sale, anything from the manors of which he is keeper; for they ought not, by right, to buy the things or take for price what they themselves ought to make profitable and sell. No bailiff shall allow any horse or aver, ox or cow, young beast, wether or ewe, or hog in his charge to be flayn before it be seen for

defaute il sont morz car chiual o auer par defaute de garde put estre peri en mot de manires o par coru as iumenz ou estre neez par tomber en fosses ou en ewe estre neez par autre cas ou la charrette chargee put reuerser e neyer [1] le chiual ou le charretter put creuer le oyl ou debriser la iambe o la quisse par vnt le chiual ou le affre est perdu e ausint de bof e de vaches e de totes autre bestes.

Le motons e les mereberbiz e les hogastres par defaute de garde pount estre morz de chins ou neez ou embleez e les motons e les mereberbiz pount estre descordez e estranglez e donkes diront les gardains ke il sont acorez de sank ou il pount estre uenduz e tuez car tot est auenue pur ce est bon de auer la ueue car lem put legirement conustre vn carcois de seson ou vne pel de seson. E si le bercher se put acquiter par deuant cely ke poy enseit de x carcoys ou de xx bous emblez ou pris [2] en la forme auant dite pur rendre les peaus de chescun moton il ad bon marche.

Item le baillif apres la tondeson deit fere uenir par deuant li toz les peaus de toz les berbiz tuez en larder ou mort de morine e donke put il ueer quant les sont de seson e queus sont eschorchez sanz garant e veue e donke deit il ver ke toz les peaus e les berbiz soient de vn merch e ke la lene de peaus soit pursiuant [3] e ke les peaus ne soient changez ne acatez e donkes vende les peaus oueke la lene e il deit la leyne vendre par saac ou par toyson solom ce ke il veit le pru e le auantage [4] greindre e si il le vende par saks chescun saac peisera xxx peres de leyne en touche [5] ou xxviij peres par pere e par balance bin peise par dreite perre de xij libr.[6] [7] E le bailliff ou aucun en ky il se afie soit chescun an al mercher e al dymer des aigneus e al dymer de la leyne e de peaus pur le ju de boute en correie.

Le bailliff deit ver e comander en aust par les maners

[1] tuwer.
[2] [a la faude.]
[3] as berbez.
[4] seyt.
[5] clos.
[6] laffres.
[7] [e demie.]

what default it died. For, from want of guard, a horse or aver may perish in many ways—by running to the mares, or be drowned by falling into ditches or water, or be hurt in some other way; or the loaded cart may overturn and hurt the horse, or the driver may put out its eye or break its leg or thigh, whereby the horse or the aver is lost. And so with oxen, and cows, and all other beasts.

The wethers and ewes and hogs, by want of guard, may be killed or hurt by dogs or stolen; and the wethers and ewes may struggle and be strangled, and then the keepers shall say that they died by violence, or they may be sold and killed, for although it is a chance for this it is good to have an inspection, for one can quickly know a fresh carcase and a fresh skin. And if the shepherd can acquit himself, before one who knows little, of ten carcases or twenty oxen stolen or taken in the way mentioned, by returning the skins, he has a good bargain.

And the bailiff ought, after shearing, to cause all the skins of all the sheep killed in the larder or dead of murrain to be brought before him, and then he can see how many are fresh and which are flayn without leave and inspection; and then he must see that all the skins of the sheep are of one mark and that the wool and the skins match, and that the skins be not changed or bought, and then sell the skins with the wool. And the wool ought to be sold by sack or by fleece, according as he shall see there is the greatest profit and advantage. And if he sell by sack, each sack shall weigh xxx stone of wool by touch, or xxviij stone by stone and balance, well weighed by the right stone of twelve pounds. And the bailiff, or some one in whom he trusts, should be every year at the selling and tithing of the lambs and at the tithing of the wool and skins, because of fraud.

The bailiff ought, in August, to see and command

ke les bleez soient nettement quilliz e siez e ouel e ke la gauele e les garbes soient petit e si put le[m les][1] ble plus tost secchir e la meyne garbe si put em meuz charger tasser e batre car il iad greignure perte en la grant garbe ke en la petite.

Le baillif deit apres la seint Johan fere trere hors touz le veuz boefs e les febles malement endentez e tote les villes vaches e le febles e les baraignes e de jouenes auers ke crestre ne valent en bin o mettre les en bone pasture pur engressir esi vaudra donkes le piur vn meillur e il deit tres foiz par an par gent ke seiuent del mestir fere ver toz les berbiz de sa baillie ce est a sauer apres la pasche pur la chaline de may e apres car donkes morent les berbiz puriz pur la chaline e toz ke lem trouc iceles pur certein esproue de tuer ij ou iij de meilleurs e tant de milueins e tant de pires ou par esproue del oyl ou del laine ke sen deperte de la pel soient venduz oue tote la leyne e autre foyz soient les veuz eles febles soient[2] tret auant la gule de aust e soient mis en bone pasture pur engresser e quant les meillurs sont ankes amende e engressi si soient vendu as macegrefs[3] si com lem put meuz car char de moton est plus coueitee e vendu adonkes ke apres le aust e tote le remeignant de creim ke ne put estre uendu adonkes seit vendu deuant la seint martin e la tirce foyz a la seint michel soient toz les berbiz treez car tot soient berbiz seins a la pasche e en may e deuant la gule de aust e apres si pount eles entre les deus festes de nostre dame par mauueise garde manger la teye de la nuele e les petiz blank limazons parunt eles porriront e moriront e pur ce vaut meuz ke lem se purueie auant de fere le pru de teles ou si non toz serront perduz.

Del office de prouost.

Le prouost deit estre elu e presente par commun assentement de tote la vile pur le meillur hosebonde

[1] Hole in the parchment.   [2] croym.   [3] [par parcels.]

throughout the manors that the corn be well gathered and reaped evenly, and that the cocks and sheaves be small, so will the corn dry the quicker; and one can load, stack, and thresh the small sheaf best, for there is greater loss in the large sheaf than in the small.

The bailiff ought, after St. John's Day, to cause all the old and feeble oxen with bad teeth to be drafted out, and all the old cows, and the weak and the barren, and the young avers that will not grow to good, and put them in good pasture to fatten, so the worst shall then be worth a better. And he ought, three times a-year, to cause all the sheep in his charge to be inspected by men who know their business—that is, to wit, after Easter, because of the disease of May, and later, for then sheep die and perish by the disease; and all that are found so, by the sure proof of killing two or three of the best, and as many of the middling, and as many of the worst, or by proof of the eye or of the wool, which separates from the skin, let them be sold with all the wool. And again, let all the old and weak be drafted out before Lammas, and let them be put in good pasture to fatten, and when the best have presently mended and are fat, let them be sold to the butchers; so can one do well, for mutton flesh is more sought after and sold then than after August; and let all the rest of the draft beasts which cannot be sold then be sold before Martinmas. And the third time, at Michaelmas, let all the sheep be drafted out; for although sheep are sound at Easter and in May and before Lammas, afterwards they can, between the two feasts of our Lady, by bad keeping, eat the web of the rime and the little white snails, from which they will sicken and die; and for this it is good to provide beforehand to make profit of such, for if not all will be lost.

The Office of Provost.

The provost ought to be elected and presented by the common consent of the township, as the best husbandman

e le moillur[1] aprour des autres e il deit ver ke toz les serianz de la curt soient matin leuez a fere lor mester e ke les charues soient par tens iointes e les terres soient bin arez e bin ensonees e bin atornees e semez de bone e nette semence solom ce ke les terres vodront porter e il deit uer ke il eit bone faude de clayes de fust sor le dimaigne estramee chescune nuyt dedenz[2] [por la terre amender].[3]

Item il deit ver ke il eit bone faude a motons e vn autre a mereberbiz e la terce a hogastres solom ce ke il ad berbiz ele gardein de motons deit auer en sa garde cccc motons si la pasture seit large ou plus esi ele seit estrete meyns e le gardain de mereberbiz deit auer ccc on large pasture e le gardain de hogastres cc e le prouost deit uer ke il soient bin gardez en pasture e en faude e en mesons. Le prouost deit uer ke les blez soient bin e nettement batuz issi ke rin ne remeigne [en] le estreim pur crestre sur les mesons ne en lor compost pur germir le testes ele croupes e les reimsailles de venture de van[4] soient mis ensemble e batuz e soyt puis vente e mys al autre e prenge garde le prouost ke nul batur ne nule venteresse ne prenge del ble pur aporter en sein ne en[5] huse ne en souler ne en burse pautenire ne en sak ne en sakelet musce pres de la grange e nul comble de ble ne seit mes receu de grange en gernir por acres fere mes de viij quarters soit pris le novime de tascurs par dreite mesure pur lacres ensi ke nul bussel ne nul demy bussel ne nul contel ne remeigne au prouost bar les bons baturs dehors la mesure auant dite car le combler e les bosseus e les demy busseus e les canteus e les remeignanz ke il vrent engerner sanz taille ou numbre engendrerent au seignur poy de pru e la semence ke remeint as champs de seme ke est porte al greignur ne deit pas autre foiz estre mesure ne taille e de ce deit le baillif fere prendre garde ke de la semence returnee e del comble de la mesure ne demy

---

[1] [gainnur e lo mellur.]
[2] [de lytere ou de fougere.]
[3] [E les jefunes auers e les vaches seynt chescune nuyt de denz pur la tere amender.]
[4] e del vanre.
[5] chause ne en.

and the best approver among them. And he must see that all the servants of the court rise in the morning to do their work, and that the ploughs be yoked in time, and the lands well ploughed and cropped, and turned over, and sown with good and clean seed, as much as they can stand. And he ought to see that there be a good fold of wooden hurdles on the demesne, strewed within every night to improve the land.

And he ought to see that he have a good fold for wethers, and another for ewes, and a third for hogs, according as there are sheep. And the keeper of the wethers ought to have in his keeping four hundred wethers if the pasture be large, or more, if it is narrow, fewer; the keeper of the ewes ought to have three hundred in large pasture; the keeper of the hogs two hundred. And the provost ought to see that they be well kept, in the pasture, in the fold, and in houses. The provost ought to see that the corn is well and cleanly threshed, so that nothing is left in the straw to grow in thatches, nor in manure to sprout. The husks, and the trampled corn, and the refuse of the winnowing, may be put together and threshed, and then winnowed and put with the other. And the provost must take care that no thresher or winnower shall take corn to carry it away in his bosom, or in tunic, or boots, or pockets, or sacks or sacklets hidden near the grange. And no comble must be allowed from the grange into the garner to make increase, but for eight quarters let a ninth be taken from the stacks by right measure for increase. Also no bushel, or half-bushel, or cantle shall remain with the provost for the good threshers beyond the said measure; for the comble, and the bushels, and the half-bushels, and the cantles, and others that they use in the garners without tally or number, bring little profit to the lord. And the seed which is left to sow the fields with, which is carried to the garner, ought not to be measured or tallied again; and the provost ought to take care of this, that by the seed returned, and the comble of the measure, nor by half-bushels or cantles

busseaus ne de canteaus porte en gernir ne seit fet damage al seignur par le prouost ou par autre car cele manire vnt il vse pur riwele generale. Ne nul baillif ne prouost ne face uente de ble ne de estor sanz garant de bref fors del creim de bestes e des berbiz ke deit estre treet hors si com est auantdit.

Nul forage del maner ne litire soyt uendu par prouost ne par autre mes le forage soit bin garde iekes lcm eit mestir de prendre le a la sustenance des auers issi ke nul ble ne seit batu par la defaute de forage e la litire e la feugire soient quilliz ensemble e getthe en chimins e en les rues pur composter fere. Item le prouost deit uer souent ke tote les bestes soient bin forages e gardez si com estre deiuent e ke il eient asez de pasture sanz surcarke de autre bestes e il deit enquere [1] ke les gardeins de tote manire de bestes ne aillent a feres ne a merchiz ne a lute [2] ne a tauerne par ont ke les bestes auant diz augent estraez sanz garde ne ne facent damage au seignur ne a autre mes il deyuent conge demander e mettre gardains en lur lius ke damage ne auenge e si damage auenge ou perte soient les amende prises des gardeins e les damages renduz. Item nul prouost neit poer de pleder nul pene [3] de nuly amercier mes ly ou le hayward ou le bedel receyvent les pleintes e facent les attachemenz e liurent al baillif. Item nul prouost ne deit mettre ne soffrir ke nul homme eit liureson [4] si il nel eit deserui ne il ne deit soffrir ke il ieit surcarke de surdeies en la dayerie ne ke eus ne aportent hors de la dayerie formage burre let ne cruddes en apeyrement de la dayerie ne en decres del formage. Item nul prouost ne remeigne outre vn an prouost si il ne soit esproue pur mut aprowant e leaus en ses fez e bon hosbonde. Chescun prouost deit chescun an [5] aconter oue son baillif e tailles les oueraignes e les custumes despendues en le maner par vnt il peussent del surplus en deners certeynement respondre sur lur aconte car autant uaut le dener de custume come de rente.

---

[1] veer.
[2] [ne a voylles.]
[3] pley ne.
[4] lyuereyson.
[5] symayne.

carried into the garner there be harm to the lord by the provost or any other, for this they do as a general rule. And no bailiff or provost shall sell corn or beast without warrant by writ, except the draft beasts and sheep, which ought to be drafted out as is aforesaid.

No forage or litter of the manor may be sold by the provost, or by another, but the forage must be well kept until it is necessary to take it for the sustenance of the beasts, that no corn may be threshed for want of forage; and let litter and ferns be gathered together and thrown in roads and paths to make manure. And the provost ought often to see that all the beasts are well provided with forage and kept as they ought to be, and that they have enough pasture without overcharge of the other beasts, and he ought to see that the keepers of all kinds of beasts do not go to fairs, or markets, or wrestling-matches, or taverns, by which the beasts aforesaid may go astray without guard, or do harm to the lord or another, but they must ask leave, and put keepers in their places that no harm may happen; and if harm or loss do come about, let the amend be taken from the keepers and the damage made good. Let no provost have power to hold pleas involving penalty or amerciament, but he or the hayward or the bedel may receive the plaints and make the attachments and deliver them to the bailiff. And no provost ought to permit or suffer any man to have his allowance if he be not deserving, nor ought he to allow any overcharge of under-dairywomen in the dairy, nor shall they carry from the dairy cheese, butter, milk, or curds, to the impoverishment of the dairy, and the decrease of cheese. Let no provost remain over a year as provost, if he be not proved most profitable and faithful in his doings, and a good husbandman. Each provost ought every year to account with his bailiff, and tally the works and customs commuted in the manor, whereby he can surely answer in money for the surplus in the account, for the money for customs is worth as much as rent.

Le baillif e le prouost deiuent souent ver tote les defautes de mesons de lur baillie e de murs e de fosses de hayes de chars de charrettes de charues de herces de faudes e de totes autre costages ensi ke lur purueance tant face ke il ne coueigne mie par lor defaute en le ouraigne de xij denirs perdre i marc car chescone chose vaut solom ce ke ele est gwiee. Item nul prouost al maner ne teigne table pur receyure les alanz e les venanz au custages le seignur sanz especial comandement[1] par bref car si ceus del ostiel le seignur veignent par les maners pur lur propre bosoignes le seignur ne les deit pas souder pur lur preu fere e si il veignent par iloques si pregnent lor costages de la garderobe le seignur auant ke il voisent nule part e pur ce nest pas mester de fere dews damages de vn fet e nul chiual[2] ne nul sergant ne garson ne autre ne seit receu a nul maner par nul baillif ne par nul prouost pur soriorner as costages le seignur sanz garant car rin sur acunte de ceus costages ne lur deit estre alowe.

Le office del hayward.

Le hayward deit estre vigerous homme apres car il deit tart e tempre espeyer e auironer e garder le boys les blez e les preez e totes autre choses ke touchent sa baillie e deit les attachemenz e les aprouemenz leaument fere e par deuant le prouost fere la deliuerance par plegges e liuerer les a son baillif apleder il deit les terres semer e estre outre les charuers e les hercers en tens del vn e del autre semence e il deit fere venir les priez e les custumers dues e usees pur fere les ouraignes ke fere deyuent. E il deit en feneison estre outre les fauchurs le leuer e le carier e en aust ensemblez les siors eles priez e les ouraignes e ver les bleez estre quilliz bin e nettement e tart e tempre garder issi ke point ne seit emble ne par auers mange ne defule e il deit tailler encontre le prouost tote la semence e les prieres e les custumes e les ouraignes ke fere deyuent en le

---

[1] [le seignur.]   [2] chyualyr.

The bailiff and provost must often see all the disrepairs of the houses in their charge, also of walls, ditches, hedges, carts, waggons, ploughs, harrows, folds, and all other costs, so that their foresight may do so much that it be not necessary through their fault to lose a mark for a matter of twelve pence; for each thing is valuable according as it is looked after. And no provost on a manor may keep table to receive goers or comers at the lord's cost, without special commandment by writ, for if those of the lord's house come to the manors on their own business, the lord need not pay them for their profit, but if they come there, let them take their expenses from the lord's wardrobe before they go anywhere, because there is no need to do two wrongs for one business. And no knight, or servant, or groom, or any other may be received on any manor by any bailiff or any provost to sojourn at the lord's expense without writ, for at the account nothing shall be allowed them for the expenses of these.

THE OFFICE OF HAYWARD.

The hayward ought to be an active and sharp man, for he must, early and late, look after and go round and keep the woods, corn, and meadows and other things belonging to his office, and he ought to make attachments and approvements faithfully, and make the delivery by pledge before the provost, and deliver them to the bailiff to be heard. And he ought to sow the lands, and be over the ploughers and harrowers at the time of each sowing. And he ought to make all the boon-tenants and customary-tenants who are bound and accustomed to come, do so, to do the work they ought to do. And in haytime he ought to be over the mowers, the making, the carrying, and in August assemble the reapers and the boon-tenants and the labourers and see that the corn be properly and cleanly gathered; and early and late watch so that nothing be stolen or eaten by beasts or spoilt. And he ought to tally with the provost all the seed, and boon-work, and customs, and labour,

maner par tot le an e¹ quei il amonte le baillif taille e aconte donkes deyuent il del remeignant sur acunte respondre.

LE OFFICE DEL SEIGNUR.²

Le seignur deit amer deu e dreiture e estre leaus e ueritable en ses diz e en ses fez e deyt hayr peche tort e mauueste. Le seignur ne se deit pas fere consailler de jouene plein de jouene sank ne de volantrif corage ke poy ou nient seyuent de fet ne de nul treslor loseingor ne de gabour ne de nul tel ke porte tesmoygne al estrenner mes il se deit consiller de prodeshommes e de leale genz e de meur age ky vnt mut veu e mut seu e ke vnt este renumee de bon los e ke vnkes ne furent de tricherie ne de nule mauueste ateinz ne conuencuz ne pur amur ne pur hange ne pur pour ne pur manace ne pur gain ne pur perte ne se fleccherreient hors de uerite pur consaillier lur seignur a escient a son damage fere.

Le seignur deit comander e ordiner ke ses acuntes soient oy chescun an mes pas en vn lyu mes par toz ses maners kar la put lem apertement sur la chose sauer e estendre le pru e le damage e il deit comander e ordiner ke nul baillif ne seit a table par les maners fors a furfet pur deners ensi ke rien ne prenge des maners fors fein busche e litire e ke nul priue estrange ne del osteil le seignur ne de aillours soit receu as maners sur le costage le seignur ne rien ne lur seit liure ne done sanz garant del bref for ce ke le baillif ou le prouost uodront aquiter de lur burses pur lur grant custages ke lem met sanz reson si com lem put ver desouz en vn³ chapitre.

Le seignur deit enquere par le sins e par autres a ses maners quant il yunt de son seneschal e de ses purueiances e de ses apruemenz fetes par ses venues en memes la manire deit il enquere des pruz e des damages⁴ par le baillif e par

---

¹ [sauer.]
² Inserted from (13) in which MS. this chapter is the first of the treatise.
³ [autre.] ⁴ [fez.]

which ought to be done in the manor throughout the year, and what it amounts to the bailiff tallies and accounts for, and they ought to answer on the account for the rest.

The office of the lord.

The lord ought to love God and justice, and be faithful and true in his sayings and doings, and he ought to hate sin and injustice, and evil-doing. The lord ought not to take counsel with young men full of young blood, and ready courage, who know little or nothing of business, nor of any juggler, flatterer, or idle talker, nor of such as bear witness by present, but he ought to take counsel with worthy and faithful men, ripe in years, who have seen much, and know much, and who are known to be of good fame, and who never were caught or convicted for treachery or any wrong-doing; nor for love, nor for hate, nor for fear, nor for menace, nor for gain, nor for loss, will turn aside from truth, and knowingly counsel their lord to do him harm.

The lord ought to command and ordain that the accounts be heard every year, but not in one place but on all the manors, for so can one quickly know everything, and understand the profit and loss. And he ought to command and ordain that no bailiff have his food in the manors except at a fixed price in money, so that he take nothing from the manors but hay, firewood, and straw; and that no friend, stranger, nor anyone from the lord's hostel or elsewhere be received at the manors at the lord's expense, nor shall anything be given or delivered to them without warrant of writ, unless the bailiff or provost wish to acquit it from their own purses for the great expense one is unnecessarily put to, as can be seen above in another chapter.

The lord ought to inquire by his own men and others on his manors as many as there are, about his seneschal and his doings, and the approvements he has made since his coming; in the same way he ought to inquire about

le prouost e quant il aura anserche del vn e del autre il deit demander ses acumturs e ses roules del acunte e donkes deit il ver e enquere ky ad bin fet e ky nun e ky ad aprue e ky non e ky ad fet pru e ky nun pru e amenusement e ceus ke il a donkes troua bons e leaus e apruanz il les retigne e face le pur quey e si nul y seit troue ke eit damage fet e nient aprue respoigne de son fet e auge adeu e si le seignur teigne les formes auant dites si put chescun seignur prodomme viure honestement e estre a son desir riches e manant sanz peche e tort a nuly ne fra.

Le seignur deit comander ke les acunturs par les maners oyent les pleintes e les torz de chescoyn ke se pleint del seneschal ou del baillif ou del prouost ou del hayward ou de autre ke del maner seit e ke pleinre dreiture set tenue a frank e a vilain costumer e as autres pleintifs si com par enqueste put estre ateint e ke les acunturs facent dreit en lur peril.

Le office des acountures.[1]

Les acunturs deiuent estre leaus sages e bin sachanz le mestir e tote les chapitres e les articles[2] al acunte en rentes en mises en issue de grange e en estor en totes autre choses ke il iapendent. Item les acuntes deyuent estre oy a chescun maner e donke porra lem sauer le pru e le damage e les fiez e les aproemenz del seneschal del baillif e del prouost e des autres car quant ke il vnt fet[3] de pru ou de damage purra estre veu par lacunte en vn jour ou en deus e donkes porra il legirement ver le sen ou la folie des auant dit senechaus baillifs prouostz e donkes pount les acunturs prendre enquestes de fez dont il sont en doutes e oir les pleintes de chescoin plentif e fere les amendes.

Le seneschal deit estre aioint as conturs ne mye com souerein ne com compaignon del acunte mes com collateral

[1] Inserted from (13).   [2] [qe apendent.]   [3] [de denz un an.]

profits and losses from the bailiff and provost, and how much he will have to seek from both. He ought to ask for his auditors and rolls of account, then he ought to see who has done well and who not, and who has made improvement and who not, and who has made profit and who not, but loss, and those he has then found good and faithful and profitable, let him keep on this account. And if anyone be found who has done harm and is by no means profitable, let him answer for his doing and take farewell. And if the lord observe these said forms, then will each lord live a good man and honestly, and be as he will rich and powerful without sin, and will do injustice to no one.

The lord ought to command the auditors on the manors to hear the plaints and wrongs of everybody who complains of the seneschal, or provost, or hayward, or any other who is of the manor, and that full justice be done to franks and vileins, customary-tenants, and other plaintiffs, such as by inquest can be had; and that the auditors do right at their peril.

The office of the auditors.

The auditors ought to be faithful and prudent, knowing their business and all the points and articles of the account in rents, in outlays, in returns of the grange and stock, and other things belonging thereto. And the accounts ought to be heard at each manor, and then one can know the profit and loss, the doings and approvements of the seneschal, bailiff, provost, and others, for as much as they have done of profit or loss can be seen by the account in a day or two, and then can soon be seen the sense or the folly of these said seneschals, bailiffs, and provosts; and then can the auditors take inquest of the doings which are doubtful and hear the plaints of each plaintiff and make the fines.

The seneschal ought to be joined with the auditors, not as head or companion of the account, but as subordinate,

car il deit respondre a les aconturs sur lacunte de ses fez e de ses comandemenz e de ses apruemenz par li fet as maners
 de ses custages bosoignables si com vn autre le baillif e le prouost soient ioint ensemble de rendre lor acunte de chescun maner plenerement de tote choses ke touchent les maners en rentes en mises en totes autre issues car nest pas reson ke le prouost ky chateus sunt au seignur e ke mut meins deit sauer par reson ke le baillif seit puni ne respoigne pur les fez le baillif desicom le baillif est a custages le seignur e a les comandemenz e les auantages e les prises[1] e est son chef e son souerein e deit le maner e le prouost par son seu e par son apruement guier e garder e toz les autres ke al maner apendent.

    Les acunturs deiuent defendre sur le acunte au baillif[2] ke nul comble de ble ne seit mes receu de grange engerner mes ke lem prenge ix quarters pur viij par certeine mesure estrike de tascurs e ke les busseus e les demy busseus e les canteus e les remeignanz ke soleient estre musee e oblie e le auantage le prouost hors de taille soit tot taille e tot acunte e[3] lautre car il vaut meuz de auer vn poy de gite de gerner ke tant perdre par an ke mult amunte par an ke ben le aperceit. E il deyuent defendre as baillifs e as prouostz ke nul chiual ne nul auer ne nul boef ne nule vache ne nul iouene auer ne nul moton ne nul mereberbiz ne nul hogastre a nul manir soit escorche sanz veue e seu par quel defaute il morut pur le peril de sus dit il ne couent pas issi dire a les aconturs sur le acunte fere par la reson de lur office car il deyuent estre si sages e si leaus e si sachant del mester ke il ne eient nul mester de autre aprise de chose ke touche le acunte.

---

[1] presenz.   [e les prouoz.]   [3] od.

for he must answer to the auditors on the account for his doings and for his commandments and approvements done by him on the manors and for necessary expenses, just as another. Let the bailiff and the provost be united to render their account of each manor fully, for all things relating to the manors in rents, outlays, and all other returns, for it is not right that the provost, who is the lord's chattel, and who reasonably must know much less than the bailiff, should be punished or answer for the doings of the bailiff, as the bailiff is in the pay of the lord and in commandments and advantages and presents is his head and superior, and ought by his sense and instruction to direct and keep the manor and the provost and all who belong to the manor.

The auditors ought on their account to the bailiff to forbid that any comble of corn be received from the grange into the garner, but that one takes nine quarters for eight by sure measure striked from the stackers, and that the bushels and half-bushels and the cantles and the rest which were wont to be hidden and forgotten, and are to the advantage of the provost if not tallied, be all tallied and all accounted for with the other, for it is much better to have a little waste of the garner than lose so much a year, which will amount to much yearly as may be clearly seen. And he must forbid through the bailiffs and provosts that any horse or aver, ox, cow, young aver, wether, ewe, or hog be in any wise flayn without inspection and knowledge of the fault by which it died, for the peril mentioned above. It is not necessary so to speak to the auditors about making audit because of their office, for they ought to be so prudent, and so faithful, and so knowing in their business, that they have no need of other teaching about things connected with the account.

Le office des charuers.

Les charuers deiuent estre genz de conisance[1] e deiuent sauer semer e les caruers e les herces depescees reperailler e amender e la tere bin gaigner e ensesoner e deyuent sauer ouelement les boefs iondre e chacer sanz ferir ne damager e il les deyuent bin forager e le forage bin garder ke il ne soit emble ne aporte e en prez e en pastures seuerales sauuement garder e autres auers ke dedenz son trouez deyuent enparker e cus e les tenurs deiuent fosser[2] en clore[3] batre e les teres remuer e fosser pur les teres achechir e les ewes asseuer ne il ne deyuent nul boef escorcher iekes lem eit veu e enqueres par quele defaute il morut ne il ne deiuent en les boueries nul feu aporter pur alumer ne pur eus memes eschaufer ne nule chandeile leinz alumer ne auer si il ne seit en lanterne e pur grant mestir e peril.

Le office des charrettirs.

Le charrenter deit estre sachant de son mester pur garder ses chivaus e nyer e pur somages e cariages fere sanz peril de ses chiuaus ke il ne seient surkarkez ne trop trauaillez sermenes ne afoles il doyt sauer amender son harnays e le atil de sa charrette. E le baillif e le prouost deiuent ver e sauer quant de foiz les charrettes pount aler le iour por marler e pur composter ou fein ou ble merin ou a buche carier sanz grant greuance e quant defoiz il pount aler le ior donkes deiuent les charretters a chif de la simayne respondre de chescune iornee. Ne nul charretter ne autre ne face escorcher chiual charretter ne auer sanz veue e comandement de son souerein ieke lem sache pur quey e par quele defaute il morut si com desus est dit. Ne nul charetter ne porte fu ne chandeile en les estables si la chandeile ne seit en lanterne e ce pur grant mester e donkes seit porte e garde

---

[1] conissance. [2] [e les hayes.] [3] [e bleez.]

THE OFFICE OF PLOUGHMEN.

The ploughmen ought to be men of intelligence, and ought to know how to sow, and how to repair and mend broken ploughs and harrows, and to till the land well, and crop it rightly; and they ought to know also how to yoke and drive the oxen, without beating or hurting them, and they ought to forage them well, and look well after the forage that it be not stolen nor carried off; and they ought to keep them safely in meadows and several pastures, and other beasts which are found therein they ought to impound. And they and the keepers must make ditches and build and remove the earth, and ditch it so that the ground may dry and the water be drained. And they must not flay any beast until some one has inspected it, and inquired by what default it died. And they must not carry fire into the byres for light, or to warm themselves, and have no candle there, or light unless it be in a lantern, and for great need and peril.

THE OFFICE OF WAGGONERS.

The waggoner ought to know his trade, to keep the horses and curry them, and to load and carry without danger to his horses, that they may not be overloaded or overworked, or overdriven, or hurt, and he must know how to mend his harness and the gear of the waggon. And the bailiff and provost ought to see and know how many times the waggoners can go in a day to carry marl or manure, or hay or corn, or timber or firewood, without great stress; and as many times as they can go in a day, the waggoners must answer for each day at the end of the week. No waggoner or other shall cause a cart-horse or aver to be flayn without inspection and the command of his superior, until it be known why and for what default it died, as is said above. And no waggoner shall carry fire or candle into the stables, unless the candle be in a lantern, and this for great need, and then it must be carried and watched by

par autre ke par ly. Chescon charretter gise chescune nuyt oue les chiuaus e prenge tele garde dont il uoudra sanz damage respondre e ensi les bouers gisent en meymes la manire oue lur boefs..

LE OFFICE LE VACHER.

Le wacher deit estre conisant de son mester sachant e bin gardant ses vaches e bin deit norrir les veaus de sa seuerison. Il deit uer ke il eit tors beaus e granz e bin aligees[1] pastures les vaches pur curre quant il vodront e ke nule vache ne treez ne lete outre le seint michil pur fere formage de regain car cel leter e cel rewain fet les vaches amegrir e afeblir e les fet curre plus tart vn autre an[2] e le let en est le meillur e la vache plus poure e il deit uer ke ses auers soient bin forages e bin gardez en yuer e en este si com il voudra[3] ne nule vache ne nul auer ne soit escorche einz ke son souerein le eit veu e sache par quele defaute il morut. Ne nul fu ne nule chandeile ne seit porte en la vacherie fors en la forme auant dite e bin face chescun an de chescune vacherie trere le ville vaches malement endentez e les baraignes e le creim de jouenes auers ke crestre[4] en bin ne volent soient[5] venduz en la forme auant dite e le vachir mette chescune nuyt les vaches e les autres auers en la faude en la seson e la faude soit bin estrame de litire ou de feugire si com desus est dit e il meymes i gise les nuyz oue ses vaches.

LE OFFICE DE PORCHIR.

Le porchir deit estre a les maners la ou les pors pount estre sustenu e garir en foreste ou en boys ou en sauagine ou en marays sans sostenance de la grange e si les porz pount estre sustenuz oue petite sustenance de la grange en

[1] en les.
[2] [e les veaus serrunt plus petitz ou plus fibles e vaudront le meyns.]
[3] respondre.
[4] teyer.
[5] [mys en bone pasture pur engresser e puys seyent.]

another than himself. Each waggoner shall sleep every
night with his horses, and keep such guard as he shall wish
to answer for without damage; and so shall the oxherds
sleep in the same way with their oxen.

THE OFFICE OF COWHERD.

The cowherd ought to be skilful, knowing his business
and keeping his cows well, and foster the calves well from
time of weaning. And he must see that he has fine bulls
and large and of good breed pastured with the cows, to mate
when they will. And that no cow be milked or suckle her
calf after Michaelmas, to make cheese of rewain; for this
milking and this rewain make the cows lose flesh and
become weak, and will make them mate later another year,
and the milk is better and the cow poorer. And he ought
to see that the avers be well supplied with forage, and well
kept in winter and summer, as he shall wish to answer, and
that no cow or aver be flayn before his superior has seen it
and known by what default it died. And no fire or candle
shall be carried into the cowhouse, except in the manner
aforesaid. And every year, from each vaccary, cause the
old cows with bad teeth, and the barren, and the draft of
the young avers that do not grow well to be sorted out that
they may be sold in the way aforesaid. And every night
the cowherd shall put the cows and other beasts in the fold
during the season, and let the fold be well strewed with
litter or fern, as is said above, and he himself shall lie
each night with his cows.

THE OFFICE OF SWINEHERD.

The swineherd ought to be on those manors where
swine can be sustained and kept in the forest, or in woods,
or waste, or in marshes, without sustenance from the
grange; and if the swine can be kept with little sustenance
from the grange during hard frost, then must a pigsty be

grant gelee a donkes soit porcherie fete en marais ou en boys ou les pors isoient nuyt e iour e donkes quant les troyes vnt purceus oue les plus febles porz soient chace as maners e soient sustenu des eschetes tant com la forte gelee dure ele mau tens e puys rechaces arire as autres [1] e si il nyad boys, mareys, ne sauagine, ou les pors se pount sustenir sanz tot estre sustenu de la grangre. Nul porcher ne pork ne seit au maner for tant solement [2] ke peussent estre sustenuz en aust del estoble e des eschetes de grange e quant lem bat les blez pur vendre e ausi tost com il sont en bon point e amendez soient venduz kar ky ke veut sustenir les pors par an tot des costages de la grange e acunter les costages a la liureison des pors e [3] del porchir ensemblement oue les damages ke il font par an de blez dews tant plus perdra ke il ne gaignera e ce put lem legirement ver ky acunter voudra.

Le office del berchir.

Chescun berchir deit trouer bon plegges de respondre de ses fez e de bin seruir e leaument tot seit il meymes compaignon al mouner e il deit couerir sa faude e treiller e amender dedenz e dehors e les treches [4] reperailler e fere e il deit gisir a la faude ly e son mastin. E il deit ses berbiz bin pestre e bin strager e bin garder ke eles ne seient morz ne detrez [5] de chens emblez ne adirez ne changer ne ke eles ne pessent les mores ne les siches ne betumees por receyure enfermetez e puriture par defaute de garde. Nul bercher ne deit departir de ses berbiz pur aler a feyres ou a marchiz ou a lutes ne a veilles ne a la tauerne sanz conge prendre ou demander e sanz bon gardain de mettre en son liu pur les berbiz garder ke nul damage ne vigne par sa defaute.

Toz les berbiz le seignur soient merche de vn merch e nule mere berbiz ne seit trete outre la feste de nostre dame [6]

[1] [maners.]
[2] [kant.]
[3] [les solz.]
[4] cleyes.
[5] ne destruz ne vourez.
[6] la Natyuite.

made in a marsh or wood, where the swine may be night and day. And then when the sows have farrowed, let them be driven with the feeble swine to the manors and kept with leavings as long as the hard frost and the bad weather last, and then driven back to the others. And if there is no wood or marsh or waste where the swine may be sustained without being altogether kept on the grange, no swineherd or swine shall be on the manor, except only such as can be kept in August on the stubble and leavings of the grange, and when the corn is threshed for sale, and as soon as they are in good condition and well, let them be sold. For whoever will keep swine for a year from the cost of the grange alone, and count the cost and the allowance for the swine and swineherd, together with the damage they do yearly to the corn, he shall lose twice as much as he shall gain, and this will soon be seen by whoever keeps account.

The Office of Shepherd.

Each shepherd ought to find good pledges to answer for his doings and for good and faithful service, although he be companion to the miller. And he must cover his fold and enclose it with hurdles and mend it within and without, and repair the hurdles and make them. And he ought to sleep in the fold, he and his dog; and he ought to pasture his sheep well, and keep them in forage, and watch them well, so that they be not killed or destroyed by dogs or stolen or lost or changed, nor let them pasture in moors or dry places or bogs, to get sickness and disease for lack of guard. No shepherd ought to leave his sheep to go to fairs, or markets, or wrestling matches, or wakes, or to the tavern, without taking leave or asking it, or without putting a good keeper in his place to keep the sheep, that no harm may arise from his fault.

Let all the lord's sheep be marked with one mark, and let no ewes be milked after the feast of our Lady,

car eles curent plus enviz [1] vn autre an e les aigneus vaudrent le meyns. E nul berbiz ne soit escorche auant ke lem le eit veu e seu par quele defaute ele morut car si la mereberbiz murt deuant tondeison si deit la pel valer vn toyson e si ele mort apres tondeison si deit le berchir respondre del aignel e de la toyson e des carcoys de seison de la pelette e si moton murt deuant tondeison lem deit respondre de vne bone pel e del carcoys si il est de seson e si il est apres tondeison lem deit respondre de la toyson e del carcoys de seison e de la pelette e ausi de hogastre. Nul aignel ne soit done ne merche ne nule leyne ne nule pel ne soit done si par deuant le baillif non bon chastriz e de bone lene soient ouekes les mereberbiz en la seison de curre de [2] mereberbiz e les motons e les hogastres soient veu iij foiz par an par gent ke seiuent del mester ele creim treet e vendu en la forme auant dite.

Le office de la daye.

La daye deit estre lele e de bon renom e nettement porter sey e deyt sauer le mester e ce ke il li apent ne ele ne deit soffrir ke suzdaye ne autre ne preignent ne hors portent let ne bure ne creyme [3] par vnt le formage soit amenuse e la dayerie emperee e le deit bin sauer fere formage e [4] saler formage ele deit sa vessele de la dayerie sauuer e garder ensi ke ne couigne mye chescun an nouele achater. E il deit sauer le iour quant ele comence de fere formage e de queyl poys e quant ele comence de fere ij [5] formages ele iour de combin e de queil peis e donkes deit le baillif e le prouost ver souent la dayerie e les formages quant il crest e decrest e le peys e ke nul damage ne seit fet en la dayerie ne nul larcin par ont le peys seit amenuse [6] e il deyuent sauer e espruer e

---

[1] tart.
[2] les.
[3] [ne cruddes].
[4] [ouwelement].
[5] duble.
[6] [e la dayerye enpeyre].

for they will mate more tardily another year, and the lambs shall be worth less; and let no sheep be flayn before it be seen and known for what fault it died, for if the ewe die before shearing then must the skin be worth a fleece, and if it die after shearing then the shepherd must answer for the lamb and the fleece and the fresh carcase with the skin. And if a wether die before shearing, he must answer for a good skin and for the carcase, if it be fresh; and if it be after shearing he must answer for the fleece and for the fresh carcase and the skin and the hog. Let no lamb be given or marked, nor any wool or skin be given, unless before the bailiff. Let good gelded sheep with good wool be with the ewes at the time of mating. Let the ewes and the wethers and the hogs be inspected three times a year by men who know their business, and the draft picked out and sold in the way aforesaid.

The office of dairymaid.

The dairymaid ought to be faithful and of good repute, and keep herself clean, and ought to know her business and all that belongs to it. She ought not to allow any under-dairymaid or another to take or carry away milk, or butter, or cream, by which the cheese shall be less and the dairy impoverished. And she ought to know well how to make cheese and salt cheese, and she ought to save and keep the vessels of the dairy, that it need not be necessary to buy new ones every year. And she ought to know the day when she begins to make cheese and of what weight, and when she begins to make two cheeses a day, of how much and of what weight, and then the bailiff and the provost ought to inspect the dairy often and the cheeses, when they increase and decrease in weight, and that no harm be done in the dairy, nor any robbery by which the weight shall be lessened. And they ought to know and prove and

ver quant des vaches funt la perre de formage e de burre e quant de mereberbiz la perre de ce meymes ke il peussent le plus seurement respondre[1] sor lor acunte. Nule vache ne seit tretee ne letee outre la seint mychil. Ne nule mereberbiz outre la feste de nostre dame[2] par la reson auantdite.

La daye deit ayder a uenter les blez quant ele put entendre e ele deit prendre garde des oues e des gelines e respondre des issues e garder e couerir le feu ke damage ne auenge par defaute de garde.

[1] sur.   [2] [la Nativite.]

CI FINIST LA SENESCHAUCIE.

see when the cows make a stone of cheese and butter, and when the ewes make a stone of the same, that they may be able the more surely to answer in the account. No cow shall be milked or suckled after Michaelmas, and no ewe after the feast of our Lady, for the reason aforesaid.

The dairymaid ought to help to winnow the corn when she can be present, and she ought to take care of the geese and hens and answer for the returns and keep and cover the fire, that no harm arise from lack of guard.

HERE ENDS THE BOOK OF THE OFFICE OF SENESCHAL.

LES REULES SEYNT ROBERD

ROBERT GROSSETESTE.

LES REULES SEYNT ROBERD.

Sci comencent les reules qe le bon Eueske de Nichole seynt Robert Groseteste fist a la Contesse de Nichole de garder e gouerner terres e ostel qy vodra ceux reules bien e bel tenir du soen demeyne porra viure e sey memes e les soens susteyner.

La premere reule enseigne coment seygnur ou dame purra sauer en chescun maner de totes ses teres par lour parceles totes ses rentes custumes usages seruages fraunchises feez e tenemenz.

En dreit de vos foreyns teres a comencement fetes purchacer le bref le rey de enquere par serement de xij franche homes en chescun maner totes les teres par lour parceles totes les rentes custumes vsages seruages fraunchises vos feez e vos tenemenz e coe seyt ententiuement enquis e loyaument par les plus leaus e les plus sachaunz des francks e vilayns e destinctement enroule issi ke voster seneschal souereyn eyt vne roule e vos vn autre enter e chescun baillif eyt coe qe pertient a sa baillie e si pleyntifs vignent a vous pur tort qe lem lour face ou demaunde vous meymes a gardez vos roules endreyt de cel maner dount le pleyntif est e solom coe responez e festes tener dreyture.

La secunde rule enseigne coment vos purrez sauoir par enqueste commune quant est en chescun maner moeble e nient moeble.

Apres festes fere sanz delay dreyt enqueste e enrouller en vn autre roule distinctement trestouz vos maners en

THE RULES OF SAINT ROBERT.

Here begin the rules that the good bishop of Lincoln, S. Robert Groseteste, made for the Countess of Lincoln to guard and govern her lands and hostel: whoever will keep these rules well will be able to live on his means, and keep himself and those belonging to him.

THE FIRST RULE TEACHES HOW A LORD OR LADY SHALL KNOW IN EACH MANOR ALL THEIR LANDS BY THEIR PARCELS, ALL THEIR RENTS, CUSTOMS, USAGES, SERVICES, FRANCHISES, FEES, AND TENEMENTS.

Touching your foreign lands; to begin with, buy the king's writ, to inquire by the oath of twelve free men in each manor all the lands by their parcels, all the rents, customs, usages, services, franchises, fees, and tenements, and let this be carefully and lawfully inquired into by the most loyal and wisest of the freeholders and villeins, and distinctly enrolled, so that your chief seneschal may have one whole roll, and you another, and let each bailiff have what belongs to his baillie. And if plaintiffs come to you for wrong that anyone has done them, or petitioning, first look yourself at the rolls of that manor to which the plaintiff belongs, and according to them give answer and maintain justice.

THE SECOND RULE TEACHES HOW YOU MAY KNOW BY COMMON INQUEST WHAT THERE IS ON EACH MANOR, MOVEABLE OR NOT MOVEABLE.

Next, cause to be made without delay right inquest, and enrol distinctly in another roll every one of your

Engeltere chescun par say quant vos purrez e quantes des carues vos auetz en chescun liu petit ou graunt e quantes vos purrez auer quantes des acres de tere gaynable quantes de pree com bien de pasture a berbitz e com bien a vacches e issi a totes maners de auers par certeyn noumbre e com bien de moeble vos auetz en chescun liu de vife auer e retenez cel roule od vous e souent regardez le premer roule e cestuy ausi qe prestement sachetz trouer coe dount aueretz afere. Trestouz vos serganz de maners seyent a certeyn soudz de deners e apres le august vos graunges seyent closes.

La tierce reule enseigne lareisounement qe le seignur ou dame deyt fere a son haut seneschal deuant acuns de tes bons amys.

Kaunt les auantdites roules e enquestes serrunt fetes e si tost com vos purrez ela pur trauail de genz ne seyt lesse apele uostre graunt seneschal deuant acune genz de ky vos affiez e luy dytes issi. Beau sirre vos veez bien ke pur mon dreyt esclarsir e pur sauer plus certeynement le estat de ma gent e de mes teres e quey ioe pusse desore enauaunt del men fere e quey lesser ioe ay fet fere ces enquestes e ces enroulemenz ore empri com a cely a ky ioe ay baille quant ke ieo ay desoutz mey agarder e gouerner e estreytement vous comaunde ke totes mes dreytures fraunchises e mes possessions neent moebles gardez enteres saunz blemure e coe qe par paresce de autres ou par tort est sustret ou amenuse des auantdites choses a tut vostre poer repelez e mes biens moebles e mes estores par honeste e dreyturele manere multipliez e leaument gardez e les issues de mes teres rentes e moeble sanz boisdye e de leal amenusement fetes venir a mey memes e a ma garderobe a despendre solom coe ke ioe quideray qe deu seyt pae e mon honur e mon pru igise par purueaunce de mey e de vous e de mes autres amys. De autre part ieo vos comaund

manors in England, each by itself, how many ploughs you have in each place, small or great, and how many you can have; how many acres of arable land, how many of meadow, how much pasture for sheep, and how much for cows, and so for all kind of beasts according to their number; and what moveables you have in each place of live stock; and keep this roll by you, and often look at the first roll, and this also that you may quickly know how to find what you ought to do. Let all your servants on the manors be set at a fixed sum of money; and after August let your granges be closed.

THE THIRD RULE TEACHES THE DISCOURSE THAT THE LORD OR LADY OUGHT TO HAVE WITH THEIR CHIEF SENESCHAL BEFORE SOME OF THY GOOD FRIENDS.

When the aforesaid rolls and inquests have been made, and as soon as you can, that the work of your people be not hindered, call your chief seneschal before any of your people in whom you trust, and speak thus to him: 'Good sir, you see plainly that to have my rights set forth clearly, and to know more surely the state of my people, and of my lands, and what I can henceforth do with what belongs to me and what leave, I have caused these inquests and enrolments to be made; now I pray you, as one to whom I have committed trust, as many as I have under me guard and govern. And strictly I command you that you keep whole and without harm all my rights, franchises, and fixed possessions, and whatever of these said things is withdrawn or diminished by the negligence of others, or by wrongdoing, replace it as far as you are able. And my moveable goods and livestock increase in an honest and right way, and keep them faithfully. The returns from my lands, rents, and moveables, without fraud, and with lawful diminution, bring to me and to my treasury to spend according as I shall direct, that God may be satisfied, and my honour and my profit preserved by the foresight of myself and you and my other friends. Further, I strictly command that neither

fermement qe vous ne nul de ues baillifs desuz uous en nule maner par torsenouses demandes ou doutes ou acheysuners ou resceyte de presenz ou de douns ne gruez ne nepenez ne ne abatez les gens qe de me tignent riches ou poures esi en nule des auantdites maneres seyent par nul greuez penes ou abatuez e par certcyne enqueste qe ieo uoil ke vous facez en vostre eyre par tut le pussez atteyndre ke couenablement le facez amender e adrescer.

La quatre reule enseigne coment seignur ou dame poet encercer son estat de hors cest adire coment il ou ele poet viure par aan de soun demeyne.

En deus maners par asine poetz enquere uostre estat cest issi comaundez fermement ke en chescun liul entrer de vos bliez seyt gette en coste al entrer en graunge la viutyme garbe de chescun manere de ble e batu par sey e mesure e par asine de cele mesure porrez asiner trestu le remenaunt en graunge e sur coe fere lo ieo ke a meillurs maners de vostre tere enueez de vostre meyne ceux de qy vous plus affiez a entre en august al entrer de blees e a cel garder qe est auantdit. E si coe ne vous plest prenez ceste manere. Comaundez vostre seneschal ke chescun an a la seynt michel ke par prodeshommes e leals e sachaunz face asiner trestuz les taas de denz graunge e de hors de chescun manere de ble quant quarteres ipoent estre e puys com bien de quarteres en semence e en serganz la tere reprendra e donkes de tut le enter e de coe qe remeynt outre la tere e outre les sergaunz escriuez la summe e de coe assecz vos despenses de vostre hostel en payn e en cerueyse. Issi veez quant quarters de ble vos volez despendre la semeyne en payn despensable com bien en aumoigne coe est si vos despendez deus quarters le jour coe sunt quatorze quarters la semeyne coe sunt seet cenz e quatorze quarters par aan. E si vous despendez chescun jour pur accrestre voster aumoigne deus quarters e demy

you nor any of your bailiffs under you in any way, by unlawful exactions, or fear, or accusations, or receipt of presents or gifts, vex or hurt or ruin those who hold of me—rich or poor; and if in any of these said ways they are by anyone vexed, hurt, or ruined, by fixed inquest which I will that you make in your eyre wherever it can be attained, quickly make amendment and redress.'

The fourth rule teaches how a lord or lady can further examine into their estate, that is to say, how he or she can live yearly of their own.

In two ways by calculation can you inquire your estate. First this, command strictly that in each place at the leading of your corn there be thrown in a measure at the entrance to the grange the eighth sheaf of each kind of corn, and let it be threshed and measured by itself. And by calculating from that measure you can calculate all the rest in the grange. And in doing this I advise you to send to the best manors of your lands those of your household in whom you place most confidence to be present in August at the leading in of the corn, and to guard it as is aforesaid. And if this does not please you, do it in this way. Command your seneschal that every year at Michaelmas he cause all the stacks of each kind of corn, within the grange and without, to be valued by prudent, faithful, and capable men, how many quarters there may be, and then how many quarters will be taken for seed and servants on the land, and then of the whole amount, and of what remains over and above the land and the servants, set the sum in writing, and according to that assign the expenses of your household in bread and ale. Also see how many quarters of corn you will spend in a week in dispensable bread, how much in alms. That is if you spend two quarters a day, that is fourteen quarters a week, that is seven hundred and fourteen quarters a year. And if to increase your alms you spend two quarters and a half every day, that is seventeen

cest sunt en la semeyne disset quarters e demy c en le an vuit cent quarters e cinkante treys e demy. E quant vos auerez ceste summe trete de la graunt summe de voster ble donkes purrez vos trere la sume de la cerueyse sicom la custume ad este la semeyne pur les braces en voster hostel e donkes pernez garde de la summe quey remeyndra a vendre e de deners del ble e de uoster rente e des issues des plais de uos courtz e de voster estor leuez vos despenses de voster quisine de vos vins de voster garderobe e de soudes des sergaunz e estretez voster estor mes en totes maners pernez garde de voster ble qe ne seyt vendu hors de seysun ne saunz mester coe est adire si vos rentes e vos autres issues poeusent suffire a despenses de voster chambre e vins e quisine lessez voster ble ester enter deskes vous eyez le auauntage del ble de vn an e nent plus ou de demy al meyns.

La quinte reule uous aprent com sagement vos deuez ouerer kaunt gardes ou chaetes vous deliuerent.

Si garde ou eschete vos deliure sanz delay maundez vos lettres a deus plus prodeshommes e les plus leaus de cel pays od vn de vos deky vous affiez ke en nule manere ne bee a tele chose auer e fetes la garde ou la eschete estendre en totes choses e enueer a vos la estente desuz lur seals e solom coe ke counseil vous dorra e vos meymes uertetz si la retenez ou donez ou a vn des vos enter ou a deus ou a treys solom coe ke plus ou meyns de ceux auerount este en voster seruice e plus trauail soffert entour vos e pur vous e tutz iours especialment deuetz regarder cele reisoun. E pur nul conseil ne seyez trop hastif de cele chose doner deskes vous seez mout certe quey coe est e quey ceo vaut.

quarters and a half in the week, and in the year eight hundred and fifty three quarters and a half. And when you have subtracted this sum from the sum total of your corn, then you can subtract the sum for ale, according as weekly custom has been for the brewing in your household. And take care of the sum which will remain from sale. And with the money from your corn, and from your rents, and from the issues of pleas in your courts, and from your stock, arrange the expenses of your kitchen and your wines and your wardrobe and the wages of servants, and subtract your stock. But on all manors take care of your corn, that it be not sold out of season nor without need; that is, if your rents and other returns will suffice for the expenses of your chamber and wines and kitchen, leave your store of corn whole until you have the advantage of the corn of another year, not more, or at the least, of half [a year].

THE FIFTH RULE TEACHES YOU HOW PRUDENTLY YOU OUGHT TO ACT WHEN WARDS OR ESCHEATS FALL TO YOU.

If a ward or escheat fall to you, at once send your letters to two of the most prudent and faithful of the country, with one of your own [people] in whom you have confidence, who in no way desire to have this thing; and cause the extent of the wardship or escheat to be made in all the things, and make them send you the extent under their seals, and according to what he who counsels you shall say and yourself direct, either keep it or give it whole to one of your people, or to two or three, according as more or fewer of them shall have been in your service, and much toil have undergone about you and for you, and you ought always especially to regard this reason. And by no advice be too hasty in giving the thing until you are most sure what it is and what it is worth.

LA SYME REULE UOUS APRENT COMENT E QUANT VOS DEUETZ COMAUNDER DE CLORE E DE OUERER VOS GRAUNGES.

Comandez voster seneschal ke vos graunges par tut seyent bien closes apres le august ke nul sergaunt ne les oure sanz especial maundement ou lettre de uous ou de lui deskes seysun seyt de batre e dount seyt enueye vn leal homme ou garsun de vous ke prenge le prouost de cel liu e vn autre de la vile leal e coe tres totes houres seyent present al ouerer des graunges e al clore al batre al venter e al liuerer al seruer par taille e pernez garde ke nul seriaunt ne baillif ne seyt receueur de deners des issues mes sul le prouost ei vn autre qe eyent dount respondre.

LA SETYME REULE UOUS APRENT COMENT VOS PURREZ SAUOYR PAR COMPARER LES ACOUNTES AS ASINES DE LA ESTENTE OU DE LA DEFAUTE DE VOS SERIAUNZ E BAILLIFS DE MANERES E DE TERES.

Al chef del an kaunt totes les acountes serrount oyes e rendues de teres e de totes issues de totes despenses e de touz les maners pernetz les roules trestouz a vous e par vn ou deus de plus priuetz e leaus ke vos eyetz trespriuement fetes comperisun des roules des acontes rendues e des roules del asine de bletz e del estor ke vos feistes apres le august auaunt e solom coe ke il se acordent verrez la entente ou la defaute de vos seriaunz e baillifs e solom coe metez amendement.

LA VTYME REULE VOS APRENT LES GENERALS COMANDEMENZ QE VOS DEUEZ FERE MEYNTEFEZ A TOTE VOSTER MEIGNE.

Amonestez trestote vostre meignee meyntefetz ke tutz iceux qe vos seruunt entendunt seruir a deu e seruir vous

THE SIXTH RULE TEACHES YOU HOW AND WHEN YOU OUGHT TO COMMAND YOUR GRANGES TO BE SHUT AND OPENED.

Command your seneschal that your granges everywhere be entirely closed after August, that no servant may open them without special command or letter from you or him until threshing-time come. And then let there be sent a faithful man, or servant whom the provost shall take from that place, and another true man from the township, and all the time let them be present at the opening of the granges, and at the close, at threshing, at winnowing, at the delivery, at the survey by tally. And take care that no servant or bailiff receive the money of the returns, but only the provost and another who shall have wherewithal to answer.

THE SEVENTH RULE TEACHES YOU HOW YOU MAY KNOW TO COMPARE THE ACCOUNTS WITH THE ESTIMATE OF THE EXTENT OR THE FAULT OF YOUR SERVANTS AND BAILIFFS OF MANORS AND LANDS.

At the end of the year when all the accounts shall have been heard and rendered of the lands, and the issues, and all expenses of all the manors, take to yourself all the rolls, and by one or two of the most intimate and faithful men that you have, make very careful comparison with the rolls of the accounts rendered, and of the rolls of the estimate of corn and stock that you made after the previous August, and according as they agree you shall see the industry or negligence of your servants and bailiffs, and according to that make amendment.

THE EIGHTH RULE TEACHES YOU THE GENERAL COMMANDMENTS THAT YOU OUGHT OFTEN TO GIVE TO YOUR HOUSEHOLD.

Exhort all your household often that all those who serve you shall know to serve God and you, faithfully and

loyaument e peniblement e pur la volente deu prefere en totes choses pur facent vester volente e voster pleysir en totes choses qe ne sunt pas encountre deu.

LA NOUIME REULE VOUS APRENT KE VOUS DEUETZ DIRE SOUENT A PETITZ E A GRAUNZ DE VOSTER MEIGNE QE TOUTZ FACENT VOS COMANDEMENS.

Dites a toutz petitz e graunz e coe souent ke plenerement prestement e volenters sauz grucer e countredit facent trestutz vos comandemens qe ne sount en countre deu.

LA DYME REULE VOUS APRENT LE ESPECIAL COMANDEMENT KE VOS DEUEZ COMANDER AL MARESCHAL DE VOSTER HOSTEL.

Comaundez a ceux qe gouerunt vostre hostel deuant tote voster meigne ke ententiue garde prengent qe tote voster meigne de denz e de hors seyt leal e penible chaste e nette honeste e profitable.

LA VNTZIME REULE VOUS APRENT QUEUX DEYUENT ESTRE RESCEU DE ESTRE DE VOSTER MEIGNE DE DENZ OU DE HORS.

Comaundez ke nul ne seyt resceu ne retenu de estre de vostre meigne de denz ne de hors si lem neyt de luy renable qridaunce ke il seyt leaus sages e penibles a cel mester a ky il est resceu e od tut coe honestes e de bon murs.

LA DOTZYME REULE VOUS APRENT QUELE ENQUESTE DEYT ESTRE FETE SOUENT PAR VOSTER COMAUNDEMENT DE VOSTER MESNEE.

Comandez ke souent e ententiuement seyt fet enqueste si il eyt nul deleaus non sachaunt ord de sey letchers medlise iueroyne nent profitable e ceux qe teus serrount trouee ou de queux tele fame surd ke il seyent fors iete de voster mesnee.

painstakingly, and for the will of God to prefer in all things to do your will and pleasure in all things that are not against God.

THE NINTH RULE TEACHES YOU WHAT YOU OUGHT TO SAY OFTEN TO SMALL AND GREAT OF YOUR HOUSEHOLD, THAT ALL DO YOUR COMMANDS.

Say to all small and great, and that often, that fully, quickly, and willingly, without grumbling and contradiction, they do all your commands that are not against God.

THE TENTH RULE TEACHES YOU THE PARTICULAR COMMAND THAT YOU OUGHT TO GIVE TO THE MARSHAL OF YOUR HOSTEL.

Command those that govern your house before all your household that they keep careful watch that all your household within and without be faithful, painstaking, chaste, clean, honest, and profitable.

THE ELEVENTH RULE TEACHES YOU WHO OUGHT TO BE RECEIVED TO BE OF YOUR HOUSEHOLD INDOORS OR WITHOUT.

Command that no one be received, or kept to be of your household indoors or without, if one has not reasonable belief of them that they are faithful, discreet, and painstaking in the office for which they are received, and withal honest and of good manners.

THE TWELFTH RULE TEACHES YOU WHAT INQUEST OUGHT OFTEN TO BE MADE IN YOUR HOUSEHOLD BY YOUR COMMANDMENT.

Command that often and carefully inquest be made if there be any disloyal, unwise, filthy in person, gluttonous, quarrelsome, drunken, unprofitable, and those who shall be found so, or of whom such report is spread, let them be turned out of your household.

LA TRESTZYME REULE VOS APRENT COMENT PAR VOSTRE COMANDEMENT PEIS SERRA TENU EN VOSTRE HOSTEL.

Comandez ke en nule manere ne seyent en vostre mesnee genz qe funt en hostel pareires descord deuisuins mes toutz serront de vn acord e de vne volente com vn quer a vn alme comandez ke totz iceux qe sunt vos seriaunz de mester seyent obeysaunt e prest a ceux qe sount vtre eux en choses qui pertenent a lour mester.

LA QUATORTZYME REULE VOS APRENT COMENT VOSTER AUMOYNE PAR VOSTER COMANDEMENT SERRA LOYAUMENT GARDE E CUILLI E SAGEMENT EN POURES DESPENDU.

Comandetz ke voster aumoyne seyt loyaument cuilla e garde ne pas enuee de la table as garsons ne pas hors de sale porte ne a sopers ne a dyners de garsons uuastroille mes fraunchement sagement e atemprement sanz tenser e batre parti a poures malades e mendinans.

LA QUINTZYME REULE VOUS APRENT COMENT VOS HOSTES DEYUENT ESTRE RESCEUZ.

Comaundetz fermement ke trestoutz les hostes séculers e religius seyent des porters des usschers de marchales prestement e curteysement e a bele chere resceu des seneschals e de tutz curteysement apele e en meme la manere herbergetz e seruitz.

LA SESTZIME REULE VOUS APRENT EN QUELE VESTURE VOS GENZ VOUS DEYUENT SERUIR A VOSTER MANGER.

Comandez ke vos chiualers e trestoutz vos gentils hommes qe vos robes pernent ke meymes ces robes chescun iour e nomement a voster manger e en voster presence usunt pur vostre honour garder ne pas veuz tabartz e soulletz herigaudz en countrefetes curtepies.

THE THIRTEENTH RULE TEACHES YOU HOW BY YOUR COMMANDMENT PEACE SHALL BE KEPT IN YOUR HOSTEL.

Command that in no way there be in your household any who make strife, discord, or divisions, in the hostel, but all shall be of one accord, of one will as of one heart and one soul. Command that all those who work at a craft be obedient and ready to those who are over them in the things which belong to their craft.

THE FOURTEENTH RULE TEACHES YOU HOW YOUR ALMS SHALL, BY YOUR COMMANDMENT, BE FAITHFULLY OBSERVED AND GATHERED, AND DISCREETLY SPENT ON THE POOR.

Command that your alms be faithfully gathered and kept, nor sent from the table to the grooms, nor carried out of the hall, either at supper or dinner, by good-for-nothing grooms; but freely, discreetly, and orderly, without dispute and strife, divided among the poor, sick, and beggars.

THE FIFTEENTH RULE TEACHES YOU HOW YOUR GUESTS OUGHT TO BE RECEIVED.

Command strictly that all your guests, secular and religious, be quickly, courteously, and with good cheer received by the seneschal from the porters, ushers, and marshals, and by all be courteously addressed and in the same way lodged and served.

THE SIXTEENTH RULE TEACHES YOU IN WHAT CLOTHES YOUR PEOPLE SHOULD WAIT ON YOU AT MEALS.

Command your knights and all your gentlemen who wear your livery, that that same livery which they use daily, especially at your meals, and in your presence, be kept for your honour, and not old tabards, and soiled herigauts, and imitation short hose.

La disethime reule vous aprent coment vous deuetz aser genz a manger en vostre hostel.

Fetes tote vostre fraunche mesnee e les hostes a plus ke lem put ser a tables de vne part e de autre ne pas ci quatre la treys e tote la frape des garsons quant la fraunche mesnee serra assise ensemble entrent e aseent e leuent. Estreytement defendez qe nule noyse ne soyt a vostre manger e vous meymes totes houres emyliu seetz de la haute table ke vostre presence a toutz ouertement cum seignur ou dame aperge a toutz e ke vous ouertement pussez de vne part e dautre veer toutz ele seruise e les defautes e a coe seyetz ententiue ke chescun iour a vostre manger eyetz ouertement deus soruewes sur voster hostel quant vous foez amanger e de coe seyetz aseure ke amerueille serretz tremutz e dote.

La disutyme reule vous aprent coment vous deuetz doner conge a vos genz ke gardent mester en voster hostel de partir en lour pays.

Al meyns ke vos poetz donetz conge a ceux qe gardent mester en voster hostel de partir en lur pays e quant vos conge donez aseez lour bref iour de reuenir a vous issi com eux vos volent seruir e si nul enparle ou gruce dites lus ke vos voletz estre seignur ou dame e ke vos voletz ke lem vos serue a vostre uoler e uostre pleyser e ke coe ne veut si vos engarnie e vos purueeretz de autres qe vos vodrount seruir a vostre pleyser dount aseez troueretz pur le vostre.

La disneuime reule vous aprent coment vostre hostel deyt estre serui al manger.

Comandetz ke voster paneter od le pain e voster botiler od la coupe ensemble pe a pe vignent deuant vos a la table auant la beneyson del manger e ke treys vadles seyent assis del mareschal chescun iour a seruir la haute table e les deus tables de coste de beyure e nul vessel de cerueyse seyt assis

THE SEVENTEENTH RULE TEACHES YOU HOW YOU OUGHT TO SEAT YOUR PEOPLE AT MEALS IN YOUR HOUSE.

Make your free men and guests sit as far as possible at tables on either side, not four here and three there. And all the crowd of grooms shall enter together when the freemen are seated, and shall sit together and rise together. And strictly forbid that any quarrelling be at your meals. And you yourself always be seated at the middle of the high table, that your presence as lord or lady may appear openly to all, and that you may plainly see on either side all the service and all the faults. And be careful of this, that each day at your meals you have two overseers over your household when you sit at meals, and of this be sure, that you shall be very much feared and reverenced.

THE EIGHTEENTH RULE TEACHES YOU HOW YOU OUGHT TO GIVE LEAVE TO YOUR PEOPLE WHO BEAR OFFICE IN YOUR HOUSE TO GO TO THEIR OWN HOME.

As little as possible give leave to those who keep office in your house to go to their own homes, and when you give leave, give them a short time to return to you, if they wish to serve you; and if any of them speak back or grumble, tell them that you will be lord or lady, and that you will that all serve your will and pleasure, and whoever will not do so send away, and get others who will serve your pleasure—of whom you will find enough.

THE NINETEENTH RULE TEACHES YOU HOW YOUR HOSTEL OUGHT TO BE SERVED AT MEALS.

Command that your panter with the bread and your butler with the cup come before you to the table, foot by foot, before grace, and that three valets be assigned by the marshal each day to serve the high table and the two tables at the side with drink. And no vessel with ale shall be

sur la table for desuz la table e le vin seyt assis sur les tables encoste soulement mes a la table la ou vous seez e vin e cerueyse seyt desutz la table for soul deuant vous seyt voster beyure sur la table. Comandetz ke voster mareschal ententif seyt de estre present sur la mesnee e nomement en sale de garder la mesnee de hors e de denz nette e sanz tenson ou noyse ou vileyne parole e ke a chescun mes apele les seruitours de aler a la quisine e il memes auge tote ueis deuant voster seneschal deske a vous e deske vostre mes seyt deuant vous assis e puys auge ester en myliu de la sale al chief e ueye ke ordeynement e saunz noyse augent les scriaunz od les mes de vne part e de autre de la sale a ceux qe serrount assignez de asseer les mes issi qe lem ne asseie ne serue desordeynement par especialte dount vous memes eyez le oyl al seruise deske les mes seyent assis en le hostel e puys entendez a voster manger e coe maundez qe voster esquele seyt issi replenie e tassee e nomement de entremes ke curteysement pusetz partir de vostre esquele a destre e a senestre par tute voster haute table e la vous plerra tut eyent eux de coe ke vous meymes auetz deuaunt vous.

LA VINTYME REULE VOUS APRENT DE PRENDRE ENSAMPLE DE SERUISE DE HOSTEL DE PRUDOME AL MANGER E AL SOPER.

E ke vous sachez le establiement del hostel le Eueske de Nichole sachez qe chescun quarter de furment rend nef vint payns de blauncks e bis ensemblement cest payn de peys de cink mars e le hostel al manger est serui de deus mes gros e pleners pur la aumoyne acrestre e de deus entremes pleners pur tote la fraunche mesnee al super de vne mes de leger chose e ausi vntremes e puys furmage e si estraunges vignent al super lem lur sert solom coo ke il vnt besoyne de plus.

placed on the table, but under the table, and wine only
shall be placed on the table; but at the table where you are,
wine and ale shall be under the table, except before you only
shall drink be on the table. Command that your marshal
be careful to be present over the household, and especially
in the hall, to keep the household within doors and without
respectable, without dispute or noise, or bad words. And at
each course call the servers to go to the kitchen, and they
themselves to go always before your seneschal as far as you
until your dishes be set before you, then go and be in the
middle of the chief hall, and see that all servants with
meats go orderly, and without noise to one part and another
of the hall to those who shall be assigned to divide the meats,
so that nothing be placed or served disorderly. Especially
do you yourself keep a watch over the service until the
meats are placed in the hostel, and then attend to your own
meals, and command that your dish be so refilled and heaped
up, and especially with the light dishes, that you may
courteously give from your dish to all the high table on the
right and on the left, and where you shall please they shall
soon have what you yourself had before you.

THE TWENTIETH RULE TEACHES YOU HOW YOU SHALL TAKE AN
EXAMPLE FROM THE SERVING AT DINNER AND SUPPER IN
THE HOUSE OF A GOOD MAN.

And know the establishment of the house of the Bishop
of Lincoln; know that each quarter of wheat shall make
nine score loaves of white and brown bread together, that
is loaves of the weight of five marks, and the hostel at meat
is served with two meats, large and full, to increase the alms,
and with two lighter dishes also full for all the freemen, and
at supper with one dish not so substantial, and also light
dishes, and then cheese. And if strangers come to supper
they shall be served with more according as they have
need.

LA VINTVNIME REULE APRENT COMENT VOTZ GENZ SE DEYUENT
AUER ENVERS VOS ESPECIALS AMYS EN VOSTRE PRESENCE E
EN VOSTRE ABSENCE.

Comandez vos chiualers e chapeleyns e serianz de mester e vos gentils hommes ke en beau semblaunt e beyte chere e prest seruise receyuent e honurent deuant vous e sanz vous en chescun liu toutz iceux qe eux pussent par vostre parole ou par vostre semblaunt aperceyure ke especialment vous sunt bien venuz e a queux vos volez especial honur ke en coe pussez especialment esprouer ke eux volent icoe ke vous voletz. E a plus ke vos porretz pur maladie ou deheit aforcez vos de manger en sale deuaunt vos genz kar seyetz certe graunt pru e honour vous enuendra.

LA VINTDEUSIME REULE VOUS APRENT COMENT VOUS DEUETZ
AUER ENVER VOS BAILLIFS E SERIAUNZ DE VOS TERES E
MANERS QUANT IL VIGNENT DEUANT VOS.

Quant vos baillifs e vos seriaunz de vos teres e maners venent deuant vous mout bel les apelez e bel parlez oueske eux e priuement e attemperment enqueretz coment vos genz bien fount e vos bliez en tere coment vos carues e voster estor se proue e teles demandes festes apertement e vostre seu serra moud le plus dote.

LA VINTETRESIME REULE VOUS APRENT DEFENDRE LES DINERS
E LES SOPERS HORS DE LA SALE.

Defendetz les diners e les sopers hors de la sale en muscettes e en chambres kar de coe surdunt de wast e nul honur a seignur ne a dame.

LA VINTE QUARTYME REULE VOS APRENT PUR QUELE RESON[1] LE
NOMBRE DES PARCELES.

Ke vous sachetz la resoun pur quey vous deuetz certeynement sauer la nombre de vos carues de tere e la nombre

[1] [vous deuez saver.]

THE TWENTY-FIRST RULE TEACHES YOU HOW YOUR PEOPLE OUGHT TO BEHAVE TOWARDS YOUR FRIENDS, BOTH IN YOUR PRESENCE AND ABSENCE.

Command that your knights, and chaplains, and servants in office, and your gentlemen, with a good manner and hearty cheer and ready service receive and honour, within your presence and without, all those in every place whom they perceive by your words or your manners to be especially dear to you, and to whom you would have special honour shown, for in so doing can they particularly show that they wish what you wish. And as far as possible for sickness or fatigue, constrain yourself to eat in the hall before your people, for this shall bring great benefit and honour to you.

THE TWENTY-SECOND RULE TEACHES YOU HOW YOU OUGHT TO BEHAVE TOWARDS YOUR BAILIFFS AND SERVANTS OF YOUR LANDS AND MANORS WHEN THEY COME BEFORE YOU.

When your bailiffs and your servants of lands and manors come before you, address them fairly and speak pleasantly to them, and discreetly and gently ask if your people do well, and how your corn is growing, and how profitable your ploughs and stock are, and make these demands openly, and your knowledge shall be much respected.

THE TWENTY-THIRD RULE TEACHES YOU TO FORBID DINNERS AND SUPPERS OUT OF THE HALL.

Forbid dinners and suppers out of the hall, in secret and in private rooms, for from this arises waste, and no honour to the lord or lady.

THE TWENTY-FOURTH RULE TEACHES YOU FOR WHAT REASON THE NUMBER OF PARCELS.

Know the reason why you ought for certainty to know the number of your ploughlands, and the number of acres

des acres de waret e de tere semee coe est ke par coe sauerez com bien de ble vous deuetz auer en gros com bien de estor com bien la tere deyt reprendre de semence dount vous deuetz sauer ke pourement respount chescune carue de tere ke ne rend cend summes de ble dount tauntes carues de tere cum vos auetz tauntes centeynes de quarters al meyntz deuetz auer v seyetz certe ke la tere est malement gaynee ou fausement semee ou le ble emblee si vous auet donke quaraunte carues de tere vos deuetz auer quatre mile quarters de ble si cinkaunte cink mile e issi auaunt vos deuetz sauer de chescun acre de waret poetz sustener par an deus berbitz almeyns dount cent acres de waret deus cent berbitz poent sustener deus cent acres quatre cent berbitz e issi auaunt si vos sauetz quauntes acres vous auetz en chescun ble asemer enqueretz com bien prent la acre de semail de cel soil de tere e contez le nombre de quarters de semence sauerez vous la issue de la semene e coe ke remeyndra.

La vintequinte reule vous aprent les deus reules de vendre e de batre voster ble.

Tenetz dous reules endreyt de vente e de batre de ble ke ia ne seyt ble vendu ke le fore ne vous remenge a estramer vos faudes de berbitz le iour e a compost de denz la court. E seyetz certe ke le estreim issi retenu vous vaudra la meyte del ble vendu toutz iours. De autre part ne soffrez en nule manere ke lem bate aueyne en nul liu deuant noel ne a prouendre ne a vente einz seez touz a achat si vous poez e apres le noel quant lem comence a semer aueine festes batre uostre aueine e cel fore batu si freschement cuntreuaudra si vn poy seyt medle od feyn trestud feyn e fore dorra greynure force a uos boefs e vigour a trauailler e bien poez entendre ke si vous volez aueyne vendre donkes la purrez vous meuz vendre e plus prendre kaunt couent ke chescun eyt a semer.

of fallow and of sown land; it is that you may know how much corn you ought to have altogether; how much stock, how much seed the land ought to yield. Know that each ploughland bears poorly that does not yield a hundred seams of corn, then of so many ploughlands as you have, so many hundreds of quarters at the least you ought to have, or be sure that the land is badly tilled, or falsely sown, or the corn stolen. If you have then forty ploughlands, you ought to have four thousand quarters of corn, if fifty, five thousand, and so on. Know that each acre of fallow ought to support yearly two sheep at the least, then a hundred acres of fallow can support two hundred sheep, two hundred acres four hundred sheep, and so on. If you know how many acres you have sown of each kind of corn, inquire how much the acre of that soil of land takes for sowing, and count the number of quarters of seed, and you shall know the return of seed, and what ought to be over.

THE TWENTY-FIFTH RULE TEACHES YOU TWO RULES FOR SELLING AND THRESHING YOUR CORN.

Observe two rules with regard to selling and threshing corn: that there be no corn sold that the straw does not remain to strew the sheepfolds daily and to make manure in the court. And be sure that the straw so kept will be always worth the half of the corn sold. For the other part do not in any wise let anyone thresh oats before Christmas, neither for provender nor for sale before all is bought, if you can, and after Christmas, when one begins to sow oats, cause your oats to be threshed, and that straw so newly threshed will be as good if a little is mixed with hay. All hay and straw give great strength to your oxen and vigour to work. And understand well that if you wish to sell oats then you shall be able to sell better and take more, when it is necessary that each may have to sow.

LA VINTESIME REULE VOUS APRENT COMENT A LA SEYNT MICHEL VOUS ORDEYNER VOSTER SOIORN DE TUT LE AN.

A la seynt michel chescun an quant vos sauerez la asine de touz vos bletz donkes purueez uostre soiorn a trestouz cel an e par quaunte semeynes a chescun liu solom les seisuns del an e les auantages del pays en char e en pesschun e en nule manere ne chargez par dette ne par longe demore les lius la uus soiournerez mes issi deuisez vos soiourns ke le liu a vostre departir ne domoerge en dette mes alchune chose remenge al meneir dunt le maner pusse surde en acres de estor e nomement en vaches e en berbitz deskes voster estor pusse aquiter vos vins vos robes vostre cire e tut vostre garderobe e coe serra en poy de tens si vous tenez e ouerez apres cest escrit sicom vous poez veer ouertement enceste manere leyne de mil berbitz en bone pasture al meynz deyt respondre de cinkaunte mars par an. Leyne de deus mil cent mars e issi auaunt countez par milleres. Leyne de mile berbitz en mesne pasture deyt al meyns rendre quaraunte mars. En grose e en feble pasture trente mars.

LA VINTE SETYME REULE VOS APRENT KE MULT VAUT LA ISSUE DES VACHES E DE BERBITZ.

Issue des vaches e des berbitz en furmage vaut amerueille de deners chescun iour en la seysun sanz veauls sanz agnels sanz le compost ke tut rend de ble e de bon.

LA VINTEUTYME REULE VOS APRENT QUELES HOURES EN LE AN VOUS DEUETZ FERE VOUS ACHAZ.

Ioe lo ke a deus seisouns del an facez vos graunt achaz coe est vos vins e vostre cire e vostre garderobe a la feire de seynt Botulf coe qe vous despenderez en Lindeseye e en Norfuke e en le val de Beuuer e en cel pays de Kauersham e en cel a suthampton de Wyncestre e somersete al Bristowe vez robes achatez a seynt yue.

THE TWENTY-SIXTH RULE TEACHES HOW AT MICHAELMAS YOU MAY ARRANGE YOUR SOJOURN FOR ALL THE YEAR.

Every year, at Michaelmas, when you know the measure of all your corn, then arrange your sojourn for the whole of that year, and for how many weeks in each place, according to the seasons of the year, and the advantages of the country in flesh and in fish, and do not in any wise burden by debt or long residence the places where you sojourn, but so arrange your sojourns that the place at your departure shall not remain in debt, but something may remain on the manor, whereby the manor can raise money from increase of stock, and especially cows and sheep, until your stock acquits your wines, robes, wax, and all your wardrobe, and that will be in a short time if you hold and act after this treatise as you can see plainly in this way. The wool of a thousand sheep in good pasture at the least ought to yield fifty marks a year, the wool of two thousand a hundred marks, and so forth, counting by thousands. The wool of a thousand sheep in scant pasture ought at the least to yield forty marks, in coarse and poor pasture thirty marks.

THE TWENTY-SEVENTH RULE TEACHES YOU HOW MUCH THE RETURN FROM COWS AND SHEEP IS WORTH.

The return from cows and sheep in cheese is worth much money every day in the season, without calves and lambs, and without the manure, which all return corn and fruit.

THE TWENTY-EIGHTH RULE TEACHES YOU AT WHAT TIMES IN THE YEAR YOU OUGHT TO MAKE YOUR PURCHASES.

I advise that at two seasons of the year you make your principal purchases, that is to say your wines, and your wax, and your wardrobe, at the fair of St. Botolph, what you shall spend in Lindsey and in Norfolk, in the Vale of Belvoir, and in the country of Caversham, and in that at Southampton for Winchester, and Somerset at Bristol; your robes purchase at St. Ives.

SUPPLEMENT

TO

LES REULES SEYNT ROBERD

BRITISH MUSEUM, SLOANE, 1986.

INCIPIUNT STATUTA FAMILIE BONE MEMORIE DOMPNI ROBERTI GROSSETEST LINCOLNIE EPISCOPI.

Let alle men be warned that seruen ȝou and warnynge be ȝcue to alle men that be of howseliolde to scrue god and ȝou trewly and diligently and to performynge or the wyllynge of god to be performed and fulfyllydde.

Fyrst let seruantis doo perfytely in alle thyngis ȝoure wylle and kepe they ȝoure commaundements after god and ryȝthwysnesse and withe oute condicioun; and also with oute gref or offense. And say ȝe that be principalle heuede or prelate to alle ȝoure seruantis both lesse and more that they doo fully reedyly and treuly with oute offense or ayenseying alle youre wille and commaundement that is not ayeynys god.

The secunde ys that ȝe commaunde them that kepe and haue kepyinge of ȝoure howseholde a fore ȝoure meynyc that bothe with in and with oute the meynyc be trewe, honest, diligent both chaste and profitabulle.

The thrydde commaunde ye that no mann be admittyd in ȝoure howse holde nother inwarde nother vtwarde but hit be trustyd and leuyd that ȝe be trewe and diligent and namely to that office to the whiche he is admyttyd. Also that he be of goode maners.

The fowreth be hit sowȝht and examined ofte tymys yf ther be ony vntrewman vnkunnynge vnhonest lecherous

stryffule drunke lewe unprofitabulle yf there be ony suche yfunde or diffamydde vppon those thyngis that they be case oute or put fro the howseholde.

The fyft commaunde ȝe that in no wyse be in the howseholde men debatefulle or stryffulle but that alle be of oonn acorde of oonn wylle euen lyke as in them ys oon mynde and oon sowle.

The sixte commaunde ȝe that all tho that seruen in ony offyce be obedient and redy to them that be a bofe them in thyngis that perteynynn to there office.

The seuenth commaunde ȝe that ȝoure gentilmen yomen and other dayly bere and were there robis in ȝoure presence and namely at the mete for ȝoure worshyppe and not oolde robis and not cordynge to the lyuery nother were they oolde schoon ne fylyd.

The viii. commaunde ȝe that ȝoure almys be kepyd & not sende not to boys and knafis nother in the alle nothe outh of the halle ne be wasted in soperys ne dyners of gromys but wisely, temperatly with oute bate or betyng be hit distribute and the departyd to powre menn beggers, sykefolke and febulle.

The ix. make ȝe ȝoure owne howseholde to sytte in the alle as much as ye mow or may at the bordis of oon parte and of the other parte and lette them sitte to gedur as mony as may not here fowre and thre there. And when youre chef maynye be sett then alle gromys may entre sitte & ryse.

The x. streytly forbede ȝe that no wyse be at ȝoure mete. And sytte ȝe euer in the myddul of the hye borde that youre fysegge and chere be schewyd to alle menn of both partyes and that ȝe may see lyȝhtly the seruicis and defawtis and diligently see ȝe that euery day in ȝoure mete seson be two men ordeyned to ouer se youre mayny and of that they shalle drede ȝou.

The xi. commaunde ȝe and yeue licence as lytul tyme as ye may with honeste to them that be in ȝoure howseholde to go home. And whenne ȝe yeue licence to them assigne ȝe to them a short day of comynge ayeyne undur peyne of

lesyng there seruice. And if ony mann speke ayen or be worthe say to hym What wille ye be Lorde ye wylle that y serue you after ʒoure wylle. And they that wylle not here that ʒe say effectually be they ywarnyd and ye shall prouide other seruantis the which shalle serue you to your wylle or plesynge.

The xii is commaunde the panytere with youre brede and the botelare with wyne and ale come to gedur afore ʒou at the tabulle afore gracys and let be there three yomen assigned to serue the hye tabulle and the two syde tabullis in solenne dayes. And ley they not the vessels deseruyng for ale and wyne uppon the tabulle but afore you. But be they layid under the tabulle.

The xiii. commaunde ye the stywarde that he be besy and diligent to kepe the maynye in hys owne persone inwarde and vtwarde and namely in the halle and at mete that they be haue them selfe honestly with out stryffe fowlespekynge and noyse. And that they that be ordeynyd to sette messys brynge them be ordre and continuelly tyl alle be serued and not inordinatly and thorow affeccion to personys or by specialte. And take ʒe hede to this tyl messys be fully sett in the halle and aftir tende ye to ʒoure mette.

The xiiij. commaunde ʒe that youre dysshe be well fyllyd and hepid and namely of entremes and of pitance with oute fat carkynge that ye may parte coureteysly to thos thatt sitte beside bothe of the ryght hande and the left throw alle the hie tabulle and to other as plesythe you thowʒght they haue of the same that ye haue. At the soper be seruantis seruid of oon messe & byʒth metis and aftir of chese. And yf the come gestis seruice schalle be haued as nedythe.

The xv. commaunde ye the officers that they admitte youre knowlechyed men familiers frendys and strangers with mery chere the whiche they knowen you to wille for to admitte and receyue and to them the whiche wylle you worschipe and they wyllenn to do that ye wylle to do that they may know them selfe to haue be welcome to ʒou and

to be welle plesyd that they be come. And al so muche as ȝe may with oute peril of sykenes & werynys ete ȝe in the halle afore ȝoure meyny. For that shalle be to ȝour profyte and worshippe.

The xvi. when your ballyfs comyn afore ȝoure speke to them fayre and gentilly in opynn place and not in priuey. And shew them mery chere & serche and axe of them how fare owre men and tenauntis & how cornys doon & cartis and of owre store how hit ys multiplyed. Axe suche thyngis openly and knowe ȝe certeynly that they wille the more drede ȝou.

The xvii. commaunde ȝe that dineris and sopers priuely in hid plase be not had & be thay forbeden that there be no suche dyners nother sopers oute of the halle for of suche comethe grete destreccion and no worship therby growythe to the Lorde.

EXPLICUNT STATUTA FAMILIE BONE MEMORIE.

GLOSSARIAL INDEX

LIST OF ABBREVIATIONS USED TO DENOTE THE AUTHORITIES REFERRED TO.

B. Burguy, G. F., *Grammaire de la Langue d'oïl*.
C. Cotgrave, R., *A Dictionarie of the French and English Tongues* (London, 1611).
D. Du Cange, *Glossarium Mediae et Infimae Latinitatis* (Paris, 1883).
G. Godefroy, F., *Dictionnaire de l'ancienne Langue française* (Paris, 1880, etc.).
H. Halliwell, J. O., *Dictionary of Archaic and Provincial Words* (London, 1850).
J. Jamieson, J., *Etymological Dictionary of the Scottish Language* (Paisley, 1879).
K. Kelham, R., *Dictionary of the Norman or Old French Language* (London, 1779).
L. Littré, E., *Dictionnaire de la Langue française* (Paris, 1863).
M. and S. Mayhew, A. L., and Skeat, W. W., *Concise Dictionary of Middle English* (Clarendon Press, 1888).
P. Palsgrave, J., *Lesclarcissement de la Langue francoyse* (London, 1530).
R. Roquefort, J. J. B., *Glossaire de la Langue romane* (Paris, 1808).
S. Skeat, W. W., *Etymological Dictionary of the English Language* (Clarendon Press, 1884).
T. Tomlins, T. E., *Law Dictionary* (London, 1835).

A

aan, 126; aun, 78, 80, year
abatez, 126; abatue, 36, 126; from abatre, to beat down, ruin, abate. B. s.v. *batre*
abosoyner, 4, to have need. G.
acatez, *see* achater
accrestre, 126; encrescent, 32, to increase. K.
acer, 60, steel. B.
achat, 62, 88, 142; achaz, 32, 62, 144, a purchase, buying
achater, 62, 116; achate, 18, 92; achatent, 2; achatet, 32; achatez, 144; acatez, 94, to buy. B.
achechir, *see* ensechit
acheysuners, 126, accusations. Cf. G. s.v. *enchoisoner*
acon, 68, 80; acoune, 64, 68, 70, 80; acun, 28, some, any. B. s.v. *alcuens*
aconte, 32, 60, 62, 66, 70, 100; acounte, 6, 32, 130; acunte, 102, 104, 106, 108, 118, an account
aconter, 100; aconte, 16, 64, 68, 104; acontent, 68; acountant, 84, to account
acontur, 86, 108; acountur, 106; acumtur, 106, an accountant

acorez, 94, killed. G. s.v. *acorer*
acoyntes, 4; acoynter, to become intimate (with one). C. s.v. *accointer*
acre, 8, 66, 68, 70, 84, 86, 90, 124, 142, an acre
acres, 98, encres, 88, increase. K.
ad, *see* auer
adeu, 106, farewell
adonc, 2 etc.; adonkes, 96 etc., then. B. II. 283
adrescer, 126, put right, redress. B. s.v. *drescer*
aers, 20, harrowed. Cf. the English, p. 49. G. s.v. *aerdre*
afere, 12 etc., to do. B.
affert, 68; afferir, to be suitable. B. I. 338
affie, 72; affiez, 126, 128; afie, 94; afiet, 16, from affier, afier, to trust, confide in. K.
afole, 110, ill-treated, wounded. B.
aforcez, 140; aforcer, to make an effort. G. s.v. *aforcier*
aignel, 78, 116; aigneus, 116; anignel, 74; aygneus, 30; aygneaus, 30, a lamb
ailors, 60; ailloure, 104, elsewhere. B.
aingnele, 78; aygnelez, 30, yeaned

aioint, 106; aioindre, to be joined. B.
alanz, 102, goers
alchune, 144, some. B. s.v. *alcuens*
alegger, 86, 90; to remove, K., discharge, C.
aler, 114; aillent, 100; auge, 106, 138; augent, 100, 138; voisent, 86, 102; voyst, 28; vount, 28, to go. B. II. 282
aleyne, 24, breath. B. s.v. *haleine*
aligees, 112: *read* alignees, as in some other copies
alme, 134, soul
alouance, 90, pay
alower, 62 etc.; alowe, 4 etc., to allow, assign. M. and S.
aloynnez, 84; aloynner, to put away, remove. K.
amaylle, *see* aumayl
amegrir, 112, to grow lean
amendement, 62, 64, 130, amendment, repairing
amender, 78 etc.; amendez, 114, to mend, amend
amendes, 86, 106; fines, C., 'yssues of a court,' P.
amenuse, 24 etc.; amenusent, 32; amenuser, 4; amenuser, to diminish, reduce. B. s.v. *menut*
amenusement, 106, 124, loss, diminution
amer, 34, 104, to love
amerciemenz, 86, amerciaments
amercier, 100; amercyetz, 4, to amerce
ameroch, 56, camomile. In the *Promptorium Parvulorum* (Camden Society) Mr. Way gives a note on Maythys: 'This plant is thus mentioned by G. de Biblesworth, Arund. MS. 220, f. 301—

"Si vous trouet en toun verger Amerokes (maþen) o gletoner (and cloten)
Les aracez de vn besagu (twybel)"

In the vocabulary of names of plants, Sloane MS. 5, is given Amarusca calida, Gall. ameroche, Aug. maithe; in another list, Sloane MS. 56, cheleye, i. mathe. The camomile is still known by the appellation Mayweed.' *Promp. Parv.* p. 321
amonestez, 130, admonish
amour, 4; amur, 104, love

amunte, 20, 108; amuntant, 20; amunter, to ascend, amount
amys, 124, friends
angoysse, 4, anguish, perplexity
anignel, *see* aignel
ankes, 96, presently. K.
anserche, 106; encercer, 126 ansercher etc., to examine, inquire into. G. s.v. *encerchier*
apayred, 40, diminished, impaired. M. and S. s.v. *apeyren*
aperceyure, 140; aperceura, 86; aperceu, 34, to perceive
aperge, 136; apereir, to be visible
apertement, 104, 140, quickly, openly. B.
apeyrement, 100, impoverishment. B. s.v. *pis*
apres, *see* aspre
aprise, 88, 90, 108, information, learning. K.
aprouement, 102, 108; apruement, 86, 88, 104, 108; aproemenz, 106; emprowement, 64; enpruement, 2, the profit from land; also improvement. T.
aprouer, 90; aprouant, 86; aprower, 2; aprowera, 18; aprowant, 100; aprue, 106; aprueys, 86; apruer, 92; apruantz, 84, 106; to augment to the utmost, to make the most of land by increasing the rent. T. s.v. *approve*
aprour, 98. 'In old statutes, bailiffs of lords in their franchises are called their approvers.' T.
apruge, 90; apruver, to approve
aquiter, 104 etc.; aquiterat, 18 acquitter, 78, to acquit
arayne, 55, spider. M. and S.
areisounement, 124, discoursing, talking with. C.
arer, 84 etc.; arez, 92; arrer, 12 etc.; arrant, 8; arre, 8; arret 12, to plough
arey, 22; behindhand. C.
arire, 114, back. K.
arrerage, 32, 34, 60, arrears
arrery, 24, delayed, frustrated. K.
arreste, 24, bearded. G. s.v. *areste*
arrue, 8; arrure, 8, 14, 18, 90; arure, 86, ploughing
arsun, 2, 64, fire
articles, 106, articles
arzilouse, 14; arzylouse, 18; arsillose, 14, clayish. C.
ascient, *see* escient

GLOSSARIAL INDEX 153

aseure, 136, assured
asine, 126, 130, 144, estimate. Cf.
G. s.v. *assene*
asiner, 126, to estimate
aspre, 14; apres, 102, sharp
asseer, 138; aser, 136; aseent, 136;
aseez, 136; assei, 138; assis,
12, 136, 138; assys, 14, to
assign, place, set, settle. B. s.v.
scoir
assentement, 96, assent
asseuer, 110, to drain. D. s.v.
assewiare. On some manors this
was a customary service; see
note on 'ad aperiendos selones
ad aquae ductum.' *Domesday of
St. Paul's* (Camden Society),
lxxix.
assise, 84, law
atamet, 12; atamer, to cut, break
up. B.
atant, 30 etc.; ataunt, 76, as
much, as many. B. II. 325
ateinz, 104, convicted. K.
atemprement, 134; attemperment,
140, moderately. B.
atil, 110, equipment, gear. G.
atorne, 92, 98; atorner, to turn,
fix
attachemenz, 100, 102, attachments
atyre, 30; atyrer, to prepare,
put in order. G. s.v. *atirer*
auantage, 20, 94, 108, 144, advantage
auditor, 62, a person appointed by
the Lord of a manor to audit
the accounts of the bailiff. *See*
Statute of Westminster, II. c. 11
auene, 84; aueine, 142; aueyne,
12, 24, 30, 142; aueyngne, 66,
74, oats
auenge, 100, 118; aueyne, 14;
auenir, to happen, occur
auentur, 14, chance, hap. C.
auer, 6 etc.; auoyr, 4 etc.; a, 4 etc.;
ad, 4 etc.; auet, 6 etc.; aueret,
10 etc.; auoyt, 2; ay, 124; eet,
10; eent, 36; eez, 4 etc.; eit,
60; eurent, 32; eyet, 4; eyent,
22; eyez, 4; eyt, 4; hauet, 20;
ount, 34; vnt, 4 etc., to have
auer, 140, to behave. G.
auer, 86, 92, 94, 98, 100, 108, 110;
affre, 8, 60, 62, 90, 94, all animals included in the stock of a
farm, but more especially horses.
D. s.v. *averia* and *afferi*. Still
used in the North for a work-
horse. H.: J.

auertitz, 10; auerty, 32, prudent.
G. s.v. *avertir*
aueryl, 12, 28, April
auge, *see* aler
August, 124, 126, 130; aust, 18, 78,
94, 96, 102, August
auironer, 102, to go round. B. s.v.
virer
aumayl, 22; amaylle, 28, great
cattle. C.
aumoigne, 126; aumoyne, 134,
alms
ausins, 88; ausint, 94, also, as
well as. K.
aust, *see* August
autresi, 64, 66, as, so. B. II. 269
axe, 149, ask
ay, *see* auer
ayceles, 30, these. B. s.v. *icel*
ayder, 118, *see* eyder
aye, 24, help, assistance. K.
ayeyne, 148; ayeynys, 147,
against; ayenseying, 147, contradiction
aygneus, 30, *see* aignel.

B

bacon, 28, bacon
baillie, 84, 86, 88, 90, 92, 102, 122,
124, the office of a bailiff
baillif, 84, 86, 88, 90, 92, 94, 96,
100, 102, 104, 106, 108, 110, 116,
122, 126, 130, 140; baylif, 6, 60,
64, 74; baylyf, 10, 32, bailiff;
chefs bailifs, 62, head-bailiff; suz
bailifs, 62, under-bailiff
balance, 94, balance; *see* touche
bar, 98: *read* par
baraigne, 96, 112; barayngne,
64, barren
bargaynnes, 92; bargaynner, to
bargain
Bartholomeus, 38, Bartholomew
Anglicus
baterie, 72, threshing
batre, 18, 62, 72, 96, 110, 130;
batu, 30, 92, 98, to thresh corn. C.
batre, 90: *read* brace
batur, 92, 98, thresher
bedel, 90, 92, 100, official sum-
moner
bee, 128; beer, to desire, wish for.
B. s.v. *baer*
bef, *see* boef
belement, 8, easily
ben, *see* bin
beneyson, 136, blessing
bens, 2 etc., goods, possessions

berbiz, 36, 37, 86, 94, 96, 98, 100, 114; berbitz, 78, 124, 142; berbetz, 28; berbys, 22; berbyz, 28, 30, sheep
bercher, 28, 30, 94; berchir, 114, shepherd
berchere, 18, 20; bercherye, 30, sheepfold
beste, 64, 100, beast
besturne, 34; besturner, to turn out of course, give a different sense to. G.
betumee, 114, bog, quagmire. G.
Beuuer, 144, Belvoir
beyueynt, 24; beyvre, to drink. B. II. 124
beyure, 136, 138, drink. K.
bez, see boef
bin, 90 etc.; ben, 92 etc., well
bis, 188, brown; see pain
blanc, 24, 26; blaunk, 74, 76, 78, dairy produce, as milk, cream, etc. G.
blank, 96; blaunc, 37; blaunck, 138, white
ble, 2, 6, 14, 18, 20, 60, 62, 66, 68, 70, 72, 76, 90, 92, 100, 102, 108, 110, 114, 118, 126, 128, 142; blee, 96, 102, 126; bletz, 130; bliez, 140, corn
blemure, 124, harm. K.
boef, 8, 86, 108, 110, 112, 142; bof, 92, 94, 96; bous, 94; befs, 12, 90; bez, buef, 10; bufs, 24, ox
boisdye, 124, fraud, deceit. B. s.v. *boisie*
bon, 144, 'meate of any frute.' P.
bosoign, 90; bosoigne, 84, 102, need, business. B. s.v. *soin*
bosoignable, 88, 108; bossognable, 88; bosoygnable, 62, necessary. B. s.v. *soin*
bosoyngnablement, 62, necessarily
bosseu, *see* bussel
botiler, 136, butler
bouche, 4, 24, mouth
bouer, 24, 112, oxherd
bouerie, 110, byre
boute en correie, fraud. Cf. G.
bowe, 12, 14, mud
boys, 6, 26, 28, 60, 66, 84, 102, 112, 114, wood
brace, 128, brewing
bref, 92, 100, 102, 104, 122, writ
bref, 136 etc., short
brenne, 49, to burn. M. and S.
brez, 72, malt. K.
Bristowe, 144, Bristol
bruere, 36, heath, K., 'whynnes,' P.
bufs, *see* boef

bure, 26, 76, 110, 116, 118; burro, 110, butter
burse, 92, 98, 104, purse
busche, 92, 104; buche, 110, underwood, K., log or great billet, C.
bussel, 12, 16, 18, 98; busseau, 100; busseu, 78, 108; bosseu, 98, bushel

C

canteu, *see* cauntel
carcoye, 94, 110; carcois, 94, carcase. K.
cardon, 12, 16, thistle. B. s.v. *chardon*
carecter, *see* charetter
cariage, 110, carting
carier, 18, 90; caiyer, 20, to carry, cart
carke, 16; karker, 16; from carker etc., to load. M. and S. s.v. *charge*
carke, 88, a load
carue, *see* charue
caruers, 110: *read* carue
cas, 90, 94, case
catillous, 4, deceitful. M. and S.
caumbre, 62, hemp. L. s.v. *chanvre*
cauntel, 16; canteus, 98, 100, 108; contel, 84, a cantle
cend, 142, hundred
centeynes, 142, hundreds. C.
ceo, *see* coe
cerreyt, *see* estre
certe, 142, sure
certeyn, 12 etc., certain
certeynte, 10, certainty
ceson, *see* seson
cestuy, 124, this
chacer, 110; chace, 28, 114, to drive. B. s.v. *chacier*
chaete, *see* eschete
chaline, 96; chalyne, 20, heat
champ, 66, 84; chaump, 66, field
chance, 6; cheance, 30; cheaunce, 2, chance
chandeile, 110, 112, candle
chapeleyn, 140, chaplain
chapitre, 92, 104, 106, chapter
chapon, 74, 76, capon
char, 28, 64, 96, flesh. B.
char, 102, cart. B.
charette, 60, 62, 94, 102, 110, cart
charetter, 62, 86, 94, 112; charrettir, 110; carecter, 20, carter
charrenter, 110, carter. C. gives *charron*, a waggon-man
charue, 6, 12, 14, 22, 60, 62, 84,

90, 92, 98, 102; **charuwe**, 22; **carue**, 124, 140, plough
charue de terre, 8; **carue de terre**, 140, 142, ploughland
charuer, 10, 14, ·62, 102, 110, ploughman
charues, 10: *read* charuer
chastriz, 116, gelded. K.
chateil, 92; **chateu**, 90, 108, chattel
chaut, 20; **chaus**, 20, hot
cheance, *see* chance
checune, 10 etc.; **chescon**, 78 etc.; **chescoin**, 84 etc.; **chescoyn**, 100; **chescun**, 12 etc.; **chun**, 84, each. B. I. 173
chef, 14, 24, 32, 62, 86, 108, 130; **chief**, 138; **chif**, 110, chief, end, head
chen, 114; **chin**, 94, dog
chere, 140, ' cheere, victuals, intertainment for the teeth.' C.
chet, *see* cheyr
cheual, 8, 10, 12, 22, 24, 60, 64; **chiual**, 86, 90, 92, 94, 108, 110, horse
cheuestres, 62, harness. B.
cheuisance, 2, 4, bargain, contract, an agreement between debtor and creditor in relation to the loan of money. K.; Ducange, s.v. *chevisautia*
cheyr, 14; **chet**, 2, 16; **chete**, 14, 28, 32; **chent**, 4; **cheunt**, 28, to fall. B. s.v. *chaor*
chimin, 92, 100, road
chimineunt, 36; **chiminer**, to walk, go
chin, *see* chen
chiualer, 134, 140; **chyualyr**, 102, knight
chyuauchent, 90; **chyuaucher**, to ride on. B. s.v. *cheval*
cinkante, 128; **cinkaunte**, 142, fifty
cire, 144, wax
cisere, 80, cider. S. s.v. *cider*
claye, 98, hurdle. C.
cler, 24, 78, clear
clerk, 60, clerk
clore, 130, to shut. B.
coe, 2 etc.; **ceo**, 2 etc., this. B. s.v. *iceo*
cok, 74, 76, cock
collateral, 106, subordinate
colouris, 49, collars
columber, 6, 88, dovehouse
comandemen, 2,86; **comandement**, 102, command
combler, 98; **comble**, 74, 98, 108; **coumble**, 16, to heap up

comoune, 70; **commune**, 86, common
compaignon, 106, 114, companion
comperisun, 130, comparison
compost, 92, 98, 142, compost, manure
composter, 90, 100, 110, to manure
concelee, 84, hidden. G.
conge, 100, 114, 136, leave
conisance, 110, intelligence
conisant, 112, skilful
conninger, 88, conygarth
conquere, 84, to maintain
conreyez, 24; **conreyer**, to curry a horse. S. s.v. *curry*
consail, 86, 88; **conseil**, 128; **counseil**, 128, counsel, advice
consailler, 104; **consiller**, 104, to counsel, advise
conscience, 4, conscience
conte, 76; **conter**, to count. B.
contel, *see* cauntel
contesse de Nichole, 122, Countess of Lincoln
conteyner, 34, to behave. B. s.v. *tenir*
contredyrunt, 26; **countredit**, 132; from **contredyre**, to contradict
contrewayter, 10; **cuntregeytet**, 4, to watch, guard against. B. s.v. *guetter*
conturs, 106, auditors
conuencuz, 104, convicted
conustre, 94, to know
copable, 64, guilty
corage, 104, courage
corde, 62, rope
corgeys, 30, pea-straw
cors, 36, body
cort, 62, 72, 74; **cor**, 80, 84; **cour**, 4, 6, 16; **court**, 128; **curt**, 20, 86, 98, court
cortilage, 6. A place adjoining the court where pot-herbs etc. were grown (*Domesday of St. Paul's*, cxxi). Kennet describes it as a pen, coop, enclosure for running. (*Parochial Antiquities*, Glossary, s.v. *curtilagium*)
coru, *see* curre
costage, 92, 102, 104, 114; **coustage**, 6; **custage**, 18, 22, 24, 102, 104, 108, cost
coste, 136; **en coste**, 26, 138, side, at the side. B. s.v. *costeit*
costume, *see* custume
costumer, *see* custumer
coture, 6, 8, 66; **conture**, 8, in Latin *cultura*, a division of land varying in quantity in different

localities. 'The nature of these Culturae is fully explained in the Extents of the manor of Swaffham and other manors belonging to Ely. The arable lands of the Demesne were divided into three Culturae, or Campi, each generally with a distinct name. . . . Each cultura or campus is described as divided into many fields of varying acreage and distinct names,' *Registrum Wigorniens*. (Camden Society), p. lxv. In the *Records of the Borough of Nottingham*, i. 405, there is mention of a cultura of land containing six selions and a gora. Spelman gives *quarentena* as an equivalent. Cf. also the text, p. 8

coue, 78, couer, to hatch. B.

coueigne, 88; couent, 60; couenant, 60; couenir, to be meet, fit. B. s.v. *venir*

coueitee, 96, from coneiter, to desire, wish for. B. s.v. *covoitous*

countrefete, 134, imitation

coupe, 136, cup. M. and S.

crache, *see* creche

creche, 12; crethe, 55; crache, 30, crib, manger, 'cratch, a rack for hay or other fodder.' Promp. Parv. 'beestes stall.' P.

creim, 100, 112, 116; croym, 96. Wares for sale, but the word is not in the Glossaries. In a Compotus printed in Hoare's *Modern Wiltshire*, I. 205, occurs *multon' de cromio*, but Hoare thinks the reading should be *cronio*, ib. p. 215; *see* also *cream*. J.

crest, 14, a crest

crester, 14; crestre, 96, 112; crest, 116; cresterunt, 28; cressaunt, 66; cressent, 20; cru, 18; creu, 18, to grow. B. II. 141

crethe : *read* creche

cretine, 36, 37; cretyne, 16, flood. D. s.v. *cretina*

creyme, 116, cream

creyous, 16, chalky

croes, 14; croez, 14; croyz, 12, full of holes. D. s.v. *crosum*

croupes, 98, the grain which has fallen on the floor of the granary. G. s.v. *crape*

croys, *see* seint

cruddes, 100, curds. M. and S.

cuilla, *see* quillir

cum, 18 etc.; as, etc. B. II. 28

cuntregeytet, *see* contrewayter

cuntreuaudra, 142; cuntrevaler, to equal in value. B. II. 111

curre, 112, 116; curent, 114; coru, 94; coure, 16; curre, etc., to run, pursue. B. I. 324

curs, 34, course

curtepie, 134, short coat or cloak. M. and S.

curteysement, 134, civilly, courteously. K.

cust, 90, cost

custage, *see* costage

custume, 6, 24, 100, 102, 122; costume, 84, 86, custom

custumer, 6, 10, 90, 102; costumer, 106; serianz de custume, 10, customary tenant

cynk, 12, 16, five

D

dae, 72, 74, 76, 78; daye, 26, 32, 116, a female servant, but in particular the one in charge of the dairy. M. and S.

daerie, 72, 74; dayerie, 88, 100, 116, dairy. M. and S.

damage, 14, 64, 66, 86, 88, 90, 100, 102, 114, 116, damage, harm

dame, 86, 122, 124, 126, 136, 140, lady

debriser, 94, to break. B. s.v. *briser*

decres, 100, decrease. K.

decrest, 110; decrestre, to decrease

dedens, 20, within

defaute, 24, 64, 94, 100, 102, 108, 112, 118, 136, fault, failing

defayle, 2; defayler, to fail

defule, 102; defuler, to spoil, ill use. B. s.v. *afoler*

degastes, 4; degaster, to waste, destroy. B. s.v. *gaster*

deheit, 140, weariness; C. s.v. *dehayte*, 'out of tune, ill at ease'

deit, *see* deuet

dekes, 78 etc.; deske, 138 etc.; deskes, 128, until. B. s.v. *dusque*

deleaus, 132, disloyal

deliuerance, 102; dyluerance, 52, deliverance

delyuerreyt, 14; delyuerer, to deliver

demesne, 6, 8, 84; dimaigne, 98, demesne

demeyne, 10, 62, 70, 122, 126; demayne, 122, own. B.

demoert, 36, 74; demore, 88, 142;

GLOSSARIAL INDEX 157

demorent, 62; demorge, 92;
demert, 30; demurge, 12, 16,
20; demurt, 30; demorer, to
remain. B.
demonstrer, 10, to show
dener, 6, 12, 24, 32, 60, 62, 86, 88,
104, 142; denir, 102, a penny
denz, 28, 36, teeth
depere, 96; depertier, to depart.
K.
depescee, 110; depiece, 12; de-
pescer etc., to break. B. s.v.
piece
deroe, 12; derree, 12, pennyworth.
C.
dereyu, 16, evil, fault. Cf. B. s.v. *roi*
descheyer, 24; descheyr, 28, to
sink, decline
descord, 134, discord
descordez, 94; descorder, to
struggle, quarrel. B. s.v. *dis-
corder*
deserui, 100; deseruir, to merit,
deserve. B. s.v. *serf*
deseyte, 18, deceit
deshireteson, 90, loss. B. s.v.
hoir gives *desheritement*, de-
pouillement, etc.
desicom, 108, as
desionz, 90; desioindre, to disunite
desir, 106, wish
desleaute, 34; desleute, 34, disloyalty
desnaturelement, 30, unnaturally
desordeynement, 138, disorderly,
unruly
desore en auant, 124, from henceforth, hereafter. K.
despendre, 30; despendent, 4;
despendu, 60, to expend, spend.
B.
despensable, *see* pain
despense, 4, 32, 34, 60, 130,
expense
destre, 34, 138, right
destruccion, 88; destruction, 6, destruction
destrure, 12; destruz, 114; detrez,
114, to destroy
desturbance, 8, 20; deturbance,
10, disturbance, hindrance
desus, 14; desuz, 14; desouz, 22,
above
desuz, 22 etc.; desouz, 84, below.
B. II. 365
dette, 142, debt
dettur, 34, debtor
deu, 2, 4, 104, 124, 130, 132, God.
deuet, 4; deusent, 34; deussent,
68; deit, 60 etc.; deiuent, 62

etc.; dey, 26; deyuent, 26;
deuer, to owe. B.
deus, 2 etc.; dews, 102, 114, two,
twice
deye, 14, finger-length
di, 37; dient, 6; dirray, 4; dirroy,
86; dye, 4; dyst, 2; dyt, 2;
dyte, 34 etc., to say, tell
dime, *see* dyme
disethime, 136, seventeenth
disneuime, 136, nineteenth
disset, 128, seventeen
disutyme, 136, eighteenth
dite, 2, a writing. B. s.v. *ditier*;
also cf. *ditty*, S.
diz, 104, sayings
doctrine, 2, doctrine
done, 4; dorra, 142; doryet, 24;
donne, 60, 80; dounent, 26;
doynnet, 4; doner, to give
donke etc., *see* dunc
dotance, 84, doubt. B. s.v. *doter*
doter, 34; dote, 136; dotet, 2, to
fear, reverence. B.
dotzime, 132, twelfth
douer, 86, to dower
doun, 126, gift
dount, 2 etc.; dont, 64 etc.; dunt,
2, by reason of, of which
doute, 126, fear. B. s.v. *doter*
drage, 70, 84, dredge. H.
drein, 60, last. K.
dreit, 84; dreite, 98, right
dreiture, 84, 104, 106; dreyture,
122, 124, right, justice
dreiturel, 90; dreyturele, 124, just
duble, 20, double
dunc, 2 etc.; donke, 84 etc.;
donkes, 94 etc., then
duree, 86; dureynt, 20; durer, to
last
duresce, 12, hardness. B. s.v. *dur*
duze, 12, twelve
dykis, 48, ditch. M. and S.
dyme, 132; dime, 78; disme, 26,
tenth, tithe
dymer, 94, to tithe
dyner, 134; diner, 140, dinner

E

ee, 68, band, troop. G. s.v. *hiee*.
See also note on *yane* in the
Durham Household Book
(Surtees Society). 'Three reapers, generally women, with a
man to bind behind them
constitute a yane, i.e. the com-
plement for one ridge'; and
as an instance of a larger number

of reapers to a piece of ground
'In shearinge wee usually sette 5
or 6 or 7 shearers to a land, but
most commonly 6 on a lande.'
Best, *Rural Economy* (Surtees
Society), p. 44.
eet, *see* auer
eez, 68, bees
einz, 112, 142, before. B.
eit, *see* auer
elire, 66; eslyre, 10; elyset, 10;
elurent, 64; elu, 96, to choose,
elect
em, 96 etc.; en, 92 etc.; om, 80,
one. B. s.v. *hons*
emble, 64; embleez, 94, from
embler, to steal. B.
emperee, 116; emperer, to become
worse. B. s.v. *pis*
empri, 124; emprier, to pray, be-
seech
emprowement, *see* aprouement
emyliu, 136, in the middle
enbust, 36; enbu, 12, 'imbued,
drunke with.' C.
encercer, *see* anserche
encheson, 4, 42, occasion. M. and S.
encupe, 86, guilty, accused. K.
endreit, 88 etc.; endreyt, 26 etc.,
with regard to. B. s.v. *droit*
enfermete, 114, sickness. B.
enfranchir, 86, to set free
engarnie, 136; engarnir, to give
notice to, warn. S. s.v. *warn*
engleis, 36; engleys, 4, English
Engletere, 124, England
engressir, 96; engresser, 96;
engressyr, 22, 28, to fatten,
grow fat. B. s.v. *cras*
enhaucer, 18; enhauce, 14, to
raise. B. s.v. *halt*
enoyter, 22, to enlarge, increase.
G. s.v. *enoitier*
enparker, 110, to impark
enparle, 136; enparler, to reason.
B. I. 310
enprent, 2, borrowing
enprente, 2; enprenter, to borrow
enpruement, *see* aprouement
enqueste, 124, inquest
enroulement, 124, enrolment
ensample, 138, example
ensauer, 28, to save
ensechit, 16; ensechys, 24;
achechir, 110; ensechir etc., to
dry
enseit, 94; ensauer, to know
ensement, 23, 30; encement, 20,
also, the same. B. s.v. *eis*
ensenser, 22, to inform. K.

ensesoner, 110; ensesonees, 86;
ensesonez, 92; ensonees, 98, to
cultivate land in due rotation.
G. s.v. *assaisonner*
ensi, 86 etc., thus
ententif, 138; ententive, 132, 136,
attentive. B.
ententinement, 122, 132, atten-
tively
enter, 122, 124, 126, 128, entire
enterement, 10, 24, entirely
entrechaufer, 36, to heat
entremes, 138, the last course of a
repast
entreweyter, 32, to watch. R.
gives *entreguetteur*, 'espion,
homme qui cherche à surprendre'
entur, 8 etc.; entour, 128, around,
about. As to the method of
ploughing described on page 8,
while the English translation
given as a supplement and that
of Lambarde translate 'entur' as
'up and down,' it has been
thought that the plough went
round a centre ridge, thus pro-
ducing curved corners to the
piece of ploughed land. See a
note by Mr. Riley in *Notes and
Queries, Second Series*, viii. 30
enviz, 116, unwillingly. B. II. 289
erge, 66: *read* orge
ert, *see* estre
erybyll, 43, arable
escarsement, 12, hardly, just. G.
eschaufer, 37, 110, to warm, heat.
B.
eschaufure, 24, inflammation. G.
eschete, 86, 114, 128; chaete, 128,
escheat
escheuuerunt, 34; eschiouz, 16;
eschenuer etc., avoid, shun. G.
s.v. *eschuir*
escient, 104; ascient, wittingly,
willingly. C.
esclarsir, 124, to explain, make
clear. C.
escorce, 30, rind or outer skin. C.
escorcher, 110; escorche, 108, 112,
115; escorchez, 94; escorces,
92, to flay, skin
escrit, 2, 60, 62, writing
escule, 28; esculer, to run off,
drain. C. s.v. *escouler*
escurement, 94, the refuse removed in the process of cleansing out. G. s.v. *escouerement*
esparnier, 62; esparnie, 22;
esparnye, 22, to spare, save,
husband. B. s.v. *espargner*

esparpylez, 20; esparpyler, to
  scatter, disperse. B.
especialte, 138; especiaute, 10,
  affection, esteem. G.
espes, 66, thick. B. s.v. *espois*
espeyer, 102, to espy
espleyter, 14, to hasten. B. s.v.
  plier
esprouer, 140; espruer, 116;
  esproue, 26, 96, 100; esprouera,
  72, to prove, verify. B. s.v. *prover*
esquele, 138, a dish. C. s.v.
  escuelle
estables, 110, stables
establisement, 138, establishment,
  household
estat, 32, 126, state
este, 12, 20, 37, 74, 76, 112,
  summer. B.
estendre, 90 etc.; estendez, 6, to
  extend, survey
estendur, 6, surveyor
estente, 2, 6, 10, extent
estoble, *see* estuble
estomak, 30, stomach
estor, 2, 6, 8, 10, 20, 22, 24, 36, 60,
  62, 72, 74, 78, 90, 100, 106, 124,
  128, 130, 142, stock or store of
  any kind
estorer, 8, to stock
estorror, cattleherd
estouers, 92, needs. B.
estraez, 100, to stray. B.
estramer, 20, 142; estrame, 30, 98,
  112, to strew or straw
estrange, 90 etc.; estraunge, 138,
  strange
estranglez, 94, strangled, stifled
estre, 8 etc.; eates, 10; ert, 8 etc.;
  cerreyt, 26; fu, 16; fust, 6 etc.;
  fussent, 20 etc.; seez, 4 etc.;
  seit, 60 etc.; serent, 62 etc.;
  sereyt, 26 etc.; serra, 12 etc.;
  serreit, 62 etc.; serret, 6 etc.;
  serreynt, 26 etc.; serrunt, 20
  etc.; seyet, 4 etc.; seyent, 4
  etc.; seynt, 26 etc.; seyt, 6 etc.;
  soient, 92; sunt, 6 etc., to be
  estre ceo, 6 etc.; besides, further.
  B. s.v. *estiers*
estreim, 98, 132, straw. B. s.v.
  estraim
estrenner, 104, present
estretez, 128; estrere, to subtract
estreyt, 8, 14, 16; estrete, 98,
  narrow. B. s.v. *estroit*
estreytement, 124, 136, straitly
estrike, 108, striked
estu, 2; estue, 2; estuier, to put
  in reserve. B. s.v. *estui*

estuble, 18, 26, 92; estoble, 114,
  stubble, haulm. B.
eueske, 122, bishop
ewe, 12, 14, 16, 24, 28, 36, 87, 84,
  94, 110, water. K.
ewouses, 16, watery
eyder, 10; eyde, 28; eydern, 26;
  ayder, 118, to help. B. s.v.
  ajude
eyet, *see* auer
eyme, 12, estimation, calculation.
  G. s.v. *esme*
eyre, 30, ground. G. s.v. *aire*
eyre, 126, eyre, circuit
eyt, *see* auer

F

familous, 30, starving. B. s.v. *faim*
fauchison, 26, mowing
fauchur, 102, mower
faude, 20, 22, 30, 36, 98, 102, 112,
  114, 142, fold
fauder, 90, to fold
fause, 88, 90, false
fausement, 142, falsely
faut, 70; faillir, to fail
feble, 96, feeble
fee, 122; fye, 90, fee, fief
fein, 86, 104, 110; fen, 92; feyn,
  30, 142, hay. B.
feirs, 92, 142; feres, 100; feyres,
  114; feyretz, 8, fairs, 'holy-
  daies, feastiuall daies' etc. C.
feisses, *see* fere
female, 24, 28, 74, female
femet, 20; femer, to manure
femme, 64, 68, 72, 86, 90, woman
feneison, 102, haymaking
fens, 18, 20, 22; feens, 18, 36,
  manure. M. and S.
fer, 60, iron
fere, 2 etc.; feisses, 76; fet, 20
  etc.; fetes, 26 etc.; facet, 18
  etc.; facent, 20 etc.; fist, 74;
  font, 2 etc.; fount, 28; freit,
  86; freyt, 14; funt, 30 etc., to
  make
ferir, 110, to beat. B.
ferme, 32, farm
ferrue, 12; ferrure, 62, shoeing
fes, 3; feez, 64; fez, 86, 90, 100,
  104; fiez, 106, doings. B.
feste, 12, 22, feast; la feste de
  nostre dame, 114, 118, Sept. 8;
  les deus festes de nostre dame,
  96, probably Sept. 8 and March 25
fet, 4, feet. M. and S.
feu, 110, 118; fu, 110, 112, fire

feuerer, 28, February
feues, 66, 70, 74, beans
feugire, 84, 92, 100, 112, fern. C.
feure, 60, smith. B.
feyne, 48, feign. M. and S.
feyntyse, 20, deceit. B. s.v. *feindre*
feyretz, *see* feire
fin, 86, 90, fine
fitz, 2; fiz, 2, son
flestrysent, 20; flestryr, to wither
folie, 106, folly. M. and S.
forage, 12, 18, 24, 30, 92, 100, 110; ferage, 12; forge, 20, straw, forage
force, 142, strength
forder, 43, to further, aid. M. and S. s.v. *forther*
fore, 142, straw. G. s.v. *fuerre*
foreste, 112, forest
foreyn, 122, outside, not belonging entirely to
formage, 26, 100, 112, 116; furmage, 26, 76, 138, 142, cheese. B. s.v. *forme*
forment, 18, 84; furment, 30, 66, 70, 74, 138, wheat. B.
fornir, 90, to bake. D. s.v. *furnare*
fors, 132, out. B.
fortune, 2, fortune
fossee, 62, 94, 102; forsse, 20, ditch
fosser, 110, to ditch
foys, 8, 28, 32, 96; foiz, 84; feys, 28, 32; fez, 74; fyez, 28, times
foyner, 28, to dig, 'wrootle,' Lamborde's translation
franc-tenan, 6, 10, free-tenant
franceys, 4, French
franche, 122; fraunche, 136, 138, free
franchise, 84, 90; fraunchise, 122, 124, franchise
franck, 122; frank, 106, frank, freeman
frape, 136, crowd, retainers. K.
fraude, 10, 16, 32, fraud
fraunchement, 134, freely
freche, 26, uncultivated ground
freit, *see* fere
freyde, 20, cold. B. s.v. *froit*
furfet, 104, contract. Brachet, *Etymological French Dictionary*, s.v. *forfait*
furmage, *see* formage
furment, *see* forment
fust, 60, 90, 98, wood. B.
fye, *see* fee
fyet, 20, 32; fyer, to trust in

fyn, 20, end
fysegge, 147, visage

G

gabour, 104, idle talker. J. s.v. *gabber*
gaignage, 92; gaynage, 2, 10, 22, tillage. B. s.v. *gagnier*
gaigner, 84, 90, 110; gaignee, 86, 143; gaynet, 18; gayneret, 22, to till. B.
gain, 104, gain
galon, 26, 76, 78, 80; galun, 26, gallon
garant, 84, 86, 90, 92, 94, warrant
garbe, 12, 94, 96, 126, sheaf
gardain, 92, 94, 98, 100, 114; gardein, 74, 98; gardeyn, 64, guardian, keeper
garde, 86, 92, 96, 118, 128, guard, watch
garderobe, 102, 128, 142, wardrobe
gardyn, 6, garden
garge, 86; garger, to guard. G.
garir, 112, to preserve, keep. B.
garok, 74, a gander
garson, 102, 134, 136; garsun, 130, groom
gast, *see* wast
gaste, 2; gastent, 4, 20, 24, 26; wastent, 30; gaster, to spoil, waste. B.
gaule, 96, a sheaf
gaynable, 124, that can be tilled
gele, 14; gelee, 114, frost
geline, 74, 76, 78, 118; gelyn, 32, hen
genice, 76, a heifer
gent, 2, 4, 92, 96, people
germir, 72, 98; germy, 14, to shoot, sprout
gerner, 16, 100, garner. P. gives 'garnier, soller a lofte.'
gerneter, 16, keeper of a garner or barn
gette, 92; gettent, 22; getthe, 100; getteront, 36; gite, 108; getter etc., to throw, cast
geu, *see* gysyr
gite, *see* gette
gloutet, 30; gloutir, to swallow
grange, 16, 18, 92, 98, 106, 108, 112, 114; graunge, 70, 72, 124, 126, 130, grange
granger, 16; graunger, 64, the keeper of a grange
grasse, 40, grease
greignur, 98, *read* grange

GLOSSARIAL INDEX 161

greignure, 96 ; greynure, 142, the greatest. B. s.v. *grant*
greindre, 88, 94, greatest. B. s.v. *grant*
gres, *see* grose
greuance, 110, pain, difficulty. B. s.v. *grief*
greyn, 70, 72, grain
grose, 142 ; gres, 72, gross
grossur, 30, greatness
grucer, 132 ; gruce, 136 ; gruez, 126, to complain, grumble
guerpi, 34 ; guerpir, to leave, quit. B.
guerre, 92, war
guesent, *see* gyser
guier, 108 ; gwier, 102 ; gwye, 10, to guide, conduct. B.
gule, 4, 96. The Gule of August, or the Feast of S. Peter ad Vincula, August 1
gyser, 28 ; gyst, 14 ; gise, 112 ; guesent, 30 ; geu, 36, to lie. B. I. 346

H

hange, 104, hatred. G. s.v. *haenge*
harnays, 110, harness
harz, *see* heroe
hastif, 128, hasty, rash
hastiuement, hastily
hauet, *see* auer
hautor, 72, height
haye, 62, 102, hedge. M. and S.
hayr, 24, 104, to dislike, hate
hayward, 84, 88, 90, 92, 100, 102, 106. A farm officer
herbage, 12, 22, herbage, pasture
herbe, 36, grass
herbergetz, 134 ; herberger, to lodge. B. s.v. *holberc*
heroe, 14, 102, 110 ; harz, 62, a harrow
heroer, 18 ; hercez, 92 ; aers, 20, to harrow
heroer, 102, harrower
herigaud, 134, an upper cloak. H.
heuede, 146, head. M. and S.
heyte, 140, lively, hearty. K. s.v. *haits*
hidle, 50, hurdle
hogastre, 92, 94, 98, 108, 114, 116, a young sheep. J. s.v. *hog*
hokeday, 32, Hockday, the second Tuesday after Easter
homage, 86, homage
homme, 64, 68, man

honur, 124, honour
hors pris, 18, except. K.
hosebonde, 96, 100, husbandman
hosebonderie, 60, 80 ; hosebondrie, 2 ; hosebandrie, 34, husbandry
hostel, 30, 126, 128, 132, 134, 136 ; ostel, 122 ; ostiel, 102, 104 ; hostel, house
houireyent, 24 ; houir, to hurt. Glossary to Lacour's *Traité inédit*
houre, 36, 78, 130, 136, 142 ; hure, 28, hour
huse, 98. 'The Housia, houicia, or house was a loose kind of garment of the cloak or mantle kind; it appears to have had sleeves and to have answered the purpose of a tunic.' — Strutt, *Dress and Habits of the People of England*, II. 364
hyde, 41, hide

I

iambe, 94, leg
iekes, 14, 32, 76, 86, 100, 110 ; ieqes, 36, 37, until, as far as. B. s.v. *dusque*
ieo, 37 ; io, 2 etc. ; ioe, 124, I. B. I. 121
igraingne, 36, spider
ihesu crist, 2, Jesus Christ. This rendering of the abbreviation used in the MS. has been adopted as being the common one, instead of 'Iesu,' the more correc form
iloques, 102 ; iluk, 32, there. B. II. 299
inhom, 66. Cf. 'Innom barley, barley sown the second crop after the ground is fallowed.'— Ray, *Glossary*, quoted by Professor Skeat in *Glossary to Fitzherbert* (English Dialect Society)
ior, 10, 14, 16, 20, 68, 74, 76 ; iour, 14, 16, 110 ; iur, 18, 24, day
iorra, 60 ; iorrer, to swear
iouene, *see* ineuene
irous, 54 ; yrrous, 28, angry. B.
irre, 28, anger
issi, 12 etc. ; issin, 70, thus. B.
issi, 4, here
.issir, 36 ; isuz, 36 ; yssir, 16. to go out. B.
issue, 10, 32, 64, 70, 72, 74, 76, 78, 106, 108, 118, 124, 128, 130, 142, issue, stock, P. ; issue de grange,

M

GLOSSARIAL INDEX

'content of thy barne.' Lambarde's translation
iuer, 16, 74, 80; yuer, 12, 14, winter
iuernage, 60, 70; yuernage, 6, 8, 92, seed sown in the autumn. G. s.v. *hivernage*
iuerner, 78, to winter
iueroyne, 132, a drunkard
iueuene, 92; iuuenes, 22; iouene, 76 etc., young. B. s.v. *juefne*
iuge, 86, judge
iumente, 64, 78, 94, mare
iustice, 86, justice

K

kalendes, 76, calends
kant, 4 etc.; kaunt, 2 etc., how many
kar, 2 etc., for
karker, 16, *see* carke
Kauersham, 144, Caversham
keue, 86, a tail. K.
ky, 64 etc., who
kynde, 40, nature. M. and S.

L

labor, 62, labour
laine, *see* leyne
lanterne, 110, lantern
larcin, 116, theft
larder, 12, 94, larder
las, 20, the
launde, 66, waste land
leal, 60 etc.; leaus, 88 etc.; lale, 32 etc., true, loyal
leat, *see* let
leaument, 60, 102, 124; loyaument, 132, loyally
leaur, 72, breadth. K.
leaute, 82, loyalty
lechent, 36; leschirunt, 24; lecher etc., to lick. B.
lee, 8, wide. B. s.v. *let*
legirement, 86 etc., easily. B.
leinz, 72, 74, 76, 110; leynz, 60, there. B. s.v. *la*
lendemeyn, 30, 76, the morrow
lentiles, 66, 'fitche corne, fatche corne.' P.
leske, 52, the groin
lesse, 43, lose. M. and S. s.v. *lesen*
lesser, 124; lesse, 124; lessez, 36, to leave
let, 24, 100, 112, 116; leat, 76, 78, milk. K.

letanz, *see* leter
letcher, 132, glutton
leter, 112; lete, 112, 118; letent, 28; letera, 24; leteres, 76, 78; letanz, to milk, give milk
lettre, 128, 130, letter
letyng, 45, hindrance, delay. M. and S.
leuyd, 146, believed. M. and S. s.v. *leuen*
leyne, 24, 28, 80, 96, 114, 142; lene, 94; laine, 96, wool
leyse, 8, width
li, 84 etc.; luy, 4 etc.; ly, 2 etc., him
limazon, 96, snail
Lindeseye, 144, Lindsey
lior, 68, binder
litire, 92, 100, 104, 112, litter
liu, 66, 92, 100, 114, 126, 130, 140, 142; lu, 62, 80; lyu, 20, 104, place
liuelode, livelihood, means of living. M. and S.
liuerer, 102, 130; liurer, 62; liueree, 86, to deliver
liuerson, 100; liureison, 114, delivery of goods, payment in kind. B.
lo, 126, 142; loer, to counsel. B.
loenge, 6, praise. B. s.v. *loer*
longement, 36, long
lor, 36 etc.; lour, 6 etc.; lur, 4 etc., their
los, 104, praise. B.
loseingor, 104, flatterer. B. s.v. *losange*
lower, 62, to award, pay
loyns, 2, far
lungure, 72, length
lus, 136, them
lute, 100, 114, wrestling. K.
lyn, 28, descent, breed. B. s.v. *lin*
lywe, 8, league. B.

M

macegref, 96, 'one who willingly buys and sells stolen flesh knowing the same to be stolen.' T.
madle, 24, 28; male, 74, male. K.
mahayne, 22, wounded, ill-treated. B. s.v. *mahain*
maladie, 34, 140; maladye, 24, sickness
malueys, *see* mauueyse
malyce, 10, malice
manace, 104, threat

GLOSSARIAL INDEX 163

manant, 106, powerful, rich. B.
s.v. *manoir*
mandement, 86, 92; maundement,
130, message, command. B.
maner, 84, 86, 88, 90, 92, 94, 100,
102, 104, 106, 108, 112, 114, 122,
126, 148; manere, 6, 60, 64,
142; meneir, 142, manor
manere, 36, 60, 68, 126; manir,
108; manire, 90, 94, 100, 104,
112, manner
manger, 134; manguent, 24, 26;
mangeuent, 90, to eat
manger, 138, dinner
marc, 102; mar, 142, mark, a unit
for computation of money, also
a coin
marchandyse, 32, merchandise
marchanz, 2, merchants
marche, 22, 72, 92, 94, bargain,
purchase
marchez, 92; marchiz, 114;
merchiz, 100, markets
mareschal, 132, 136, 138; marchals, 134, marshal
mareys, 16, 26, 114; marays, 112;
marrey, 66, marsh, bog
mariage, 86, marriage
marler, 20, 30, 90, 110; marle,
20, to marl ground
marz, 14, 20, 28, March
mascher, 30; masche, 30; mascherunt, 30, to masticate
mastin, 114, mastif
matin, 4, 90, 98; matyn, 8, morning
matinee, 28, morning
mau, 114, ill, bad. B.
mauueste, 86, 90, 104, mischief,
ill-doing. B. s.v. *malvais*
mauueyse, 12 etc.; mauueise, 36
etc.; malueys, 22, bad. B. s.v.
malvais
may, 12, 76, 90, May
mayle, 12, 18, 26, a halfpenny. B.
mayn, 4, hand
maystyng, 54, acorns and other
food the swine found in the
forest. S. s.v. *mast*
medecine, 86, medicine
medler, 18; medle, 20, 22, 30, 70,
142, to mix
medlise, 132, quarrelsome
megres, 22, spare, meagre
meignee, 130, *see* mesnee
meled, 55, mixed. M. and S. s.v.
mell
men, 124, my. K.
mendement, 90, amendment
mendinan, 134, beggar. K.
mendre, 88, smallest

merch, 94, 114; merche, 114, mark
mercher, 28, 94, to mark
mereberbiz, 26, 92, 94, 98, 108,
116, 118; mereberbit, 76, 78,
ewe
merin, 60, 62, 110, timber
mes, 138, mess, course of dishes
at table
mes, 12 etc.; mays, 30, but
mes, 16, 64, 68, 98, 108, more. G.
s.v. *mais*
meschance, 2; meschaunce, 64;
mescheance, 2, mischance
meseyse, 2, misery
mesfesanz, 88, evildeeds. B. s.v.
mesfaire
mesnee, 132, 134, 136, 138;
meignee, 130, 132; meyne, 126,
household, household servants
meson, 24, 36, 92, 98, 102; mesun,
18, 22, 30, 62, house
messer, 10, 24, 64, *see* Lambarde's
note, quoted in the Introduction,
p. xxxvi. Also cf. ' Memorandum,
quod omnes prædicti, qui tenent
tenementa cum dimidia virgata
terræ, non portabunt officium
præpositi vel ballivi, sed erit
messor, Anglice tethingman '—
Extent, printed in Scrope's *History of Castle Combe*, p. 214
mester, 6, 10, 16, 62, 88, 92, 98,
102, 108, 112, 116, 128, 132, 136;
mestir, 16, 18, 96, 106, 110, need,
trade, etc. B. s.v. *mestier*
mestilon, 70, mixtelyn
mestre, 60, master. K.
met, 104; mys, 6 etc.; metre, to
put. B.
meur, 104, ripe, mature. B. s.v.
maur
meus, 30; meux, 34; meuz, 20 etc.;
mieuz, 60; myeuz, 64 etc., better
mey, 6: *read* meyns
mey, 124, my
meye, 72, mow. G. s.v. *moie*,
quotes Walter de Biblesworth:
' une moye est dite en graunge,
e tas hors de la grange.'
meylurs, 82, better. B. s.v. *mialdres*
meyns, 6 etc.; meins, 86 etc., less
meyntefez, 133, many times. B. I.
178
meyntener, 2, to maintain
meyntz, 142, least
meyte, *see* moite
miel, 80, honey
millere, 142, thousand
miluein, 96, of middling quality. K.

M 2

mises, 4, 88, 106, 108, outlay, expense. C.
moeble, 122, 124, moveable. C. s.v. *meuble*
moert, morent, 64, 96; morte, 64; morz, 28; murge, 2, 28; murt, 114; mourir etc., to die. B. I. 361
moillong, 92, stack, cock. C. s.v. *moulon*
moistous, 36, moist
moite, 76; moyte, 74; meyte, 8, 14, half
more, 36, 66, 114, moor
morine, 28, 64, 94, murrain
mostre, see mustrer
mot, 4 etc.; moud, 140; mout, 8 etc.; mult, 108 etc.; mut, 12 etc., very, much. B. II. 308
moton, 92, 94, 98, 108, 116; motoun, 30; motun, 28, 32; mutun, 30, wether, mutton
mouner, 113, miller
moustruy, 8, see mustrer
moy, 24, month
moysture, 16, moisture
munder, 16, to weed, clean. B. s.v. *monde*. C.
mur, 62, 102, 132, wall
musce, 98; musee, 108; muscer etc., to hide. B. s.v. *mucer*
muscette, 140, secret. K.
mustrer, 20; mustre, 86; moustruy, 8; mustray, 8; mostre, 84, to show. B. s.v. *mostrer*
mydi, 28, midday
mye, 6 etc., not; ne mye pur ceo, 32, nevertheless. K.
mykell, 50, great. J. s.v. *mekil*
myster, 57, need, want. M. and S.
myuueyn, 32, of middling quality

N

namely, 147, chiefly, especially. M. and S.
naturelment, 20, naturally
nef, 138, nine
nefime, see nouime
neire, 36, black
neiz, 80, nuts
nent, 2 etc.; nient, 104 etc.; nyent, 64 etc., nothing. B. s.v. *neant*
nettireyt, 14; nettir, to clean. G.
neyer, 94; nesz, 94, to drown, endamage. K.
Nichole, 122, 138, Lincoln
noel, 142; nouel, 72, 76, Christmas
non, 2, 66; noun, 34, name. B. s.v. *nom*

noreture, 28; nurture, breeding B. s.v. *norir*
Norfuke, 144, Norfolk
norrir, 112; noryr, 20; noriz, 16; noryt, 20, to nourish
nostre, 96 etc., our
nouime, 132; novime, 98; nefime, 16, ninth
noument, 34; noumer, to name
noune, 8, 28, 36, the ninth hour of the day: that is, three o'clock, afternoon
noyse, 136, 138, a quarrel, brawl
nuele, 96, fog, mist. G. s.v. *nuele*
nun, 106 etc., not
nus, 2, us
nuyt, 30, 98, 112; nut, 12, night
nyer, 110, to clean. G. s.v. *nier*

O

o, 18: *read* e
o, 24 etc.; od, 10 etc.; oue, 20 etc., with
o, 94 etc., or
obeysaunt, 134, obedient
oblie, 100; oblier, to forget
oef, 74, 76, egg
office, 62, office
oile, 80, oil
oir, 106; oy, 16, 104, 106; oyent, 106; oyes, 136; oyr, to hear. B. I. 371
oison, 74; osion, 76, gosling
oit, 90: *read* sit
ord, 132, 'filthie, nastie, foul.' C.
ordeynement, 32, 138, order, disposition. B.
ordinance, 4, ordinance
ordiner, 6, to ordain, order
orge, 20, 24, 66, 70, 74, 84, barley
oriace, 74, barley-straw. G. s.v. *orjas*
oste, 26; ostet, 8; ostez, 37; ouste, 20; oster, to take away
ostel, see hostel
oue, see owe
oueke, 28 etc.; ouekes, 116 etc.; oueske, 140, with. B. II. 344
ouel, 96; owel, 14, equal. G. s.v. *ivel*
ouelement, 70, 72, 76, 110, equally
ouerable, 68, 72, destined for work. G.
ouerayne, 10, 20, 22; oueraynne, 10; ourayngne, 60, 62; oueraigne, 100, 102, work
ouerer, 130; oure, 130, to open
ouerour, 30, labourer

GLOSSARIAL INDEX

ouerset, 46, 'overysettyn, or ovyr comyn.' *Promp. Parv.*
ourer, 62; ourir, 62, to work
owe, 32, 74, 76, 78; oue, 118, goose. B.
oyl, 94, 96, 138, eye

P

paer, 66, 74; pae, 124; paent, 64, 66, to pay, satisfy. B.
pain, 136; payn, 138, bread, loaf; payne de blauncks, 138, white bread; payn de bis, 138, brown bread; payn dispensable, the bread given to the servants. See note on 'forma liberationis' *Reg. Wig.* ciii.
paneter, 136, the panter, or officer of the pantry
parare, 8; pararrer, to plough thoroughly
parauenture, 36, perhaps
parcele, 122, 140, parcel, division of land
pareires, 134, strife; another reading is parcies. Cf. D. s.v. *parcier*
parente, 10, relations
paresce, 124, idleness
parfount, 12, 20, deep
pari, 60; parir, to appear. B.
park, 88, park
parker, 62, keeper of a park
parole, 138, word
pas, 8, 10, pace. M. and S.
pasche, 22, 26, 28, 96, Easter
pasture, 24, 26, 28, 36, 37, 66, 84, 90, 96, 98, 100, 124, pasture; pasture severale, 6, 86, 110, pasture land separated or enclosed
pautenire, 98, a purse. D. s.v. *pantonarius*
payle, 12, 24, straw
pays, 60, 68, 136, country
peaus, 28, 30, 64, 94; peus, 28, skins
peche, 104, sin
pee, 12, 68, 72; pe, 8, 10, 136; piez, 70; pye, 8, 14, foot
peis, 134, peace. K.
peis, 116; peise, 94; peys, 32, 116, 138; peyse, 26, 78, weight, wey. *See* Assize of Measures (*Statutes of the Realm*)
peisent, 78; peisera, 94; poyse, 26; peiser etc., to weigh
pel, 24, 94, 96, 114; peil, 62, skin, hair. B.

pelette, 114, skin
pene, 100, penalty, forfeiture. C. s.v. *peine*
penez, 126; penerunt, 34; pener, to take pains. B. s.v. *poene*
pentechoste, 22, 28; penthechoste, 22, Whitsunday
perche, 8, 66, 90, perch
pere, 2, father
pere, 76, 78, 94; perre, 94, 118, stone
perilouses, 78, perilous
pernez, *see* prendre
perouse, 2, 14, stony
pers, 4, peers
persaunt, 36, *read* pessaunt
perser, 30, to tear up, cut up. B.
perte, 86, 96, 100, 104, loss
pertient, 122; pertint, 84; pertinir, to belong to. B. s.v. *tenir*
pesaz, 30, peapods. D.
pesschun, 142, fish. K.
pestre, 30, 114; pessant, 28, 36; pessaunt, 36; pessent, 90, 114; peuz, 26, to pasture. B. II. 188
peyne, 8, trouble
pinfaude, 92, pinfold
piur, 96, worst. B.
plai, 128, plea
pleder, 90, 100, to hold plea. K.
plegge, 90, 102, 114, pledge
pleinre, 106, full
pleinte, 100, 106, plaint
pleintif, 106; plentif, 106; pleyntif, 122, plaintiff
plenerement, 10, 108, 132, fully
pleynement, 64, 74, fully
pleyser, 136; pleysir, 133, pleasure
pluous, 24, rainy
pluye, 14, rain
poer, 84, 86, 124; pouer, 20; power, 28, power, strength
poet, 2 etc.; poeit, 64; poent, 2 etc.; poez, 6; pora, 6; porra, 12 etc.; porret, 6; pount, 68; peust, 28; puet, 14; puent, 4; puisse, 60; puissent, 36 etc.; purra, 64 etc.; purrer, 68 etc.; purret, 4 etc.; purrunt, 6; pusset, 4 etc.; pussent, 16 etc.; pust, 8 etc.; put, 8 etc.; povir etc., to be able
poi, 90, little
point, 90, 114; poyn, 2; poynt, 12, 'araye, condicion, case.' P.
pokkes, 36, pocks. M. and S.
polene, 78; poleyn, 64, colt, foal
pomelyere, 53, disease of the lungs
pomun, 24, lung

166 GLOSSARIAL INDEX

porceler, 28, to farrow
porcher, 114 ; porchir, 112, 114, swineherd
porcherie, 114, swine-stye
porriront, *see* purryst
pors, 28, 86, 112, 114 ; pork, 114, pigs
porter, 134, porter
portur, 76, bearing
possession, 124, possession
potel, 26, a pottle, two quarts
pouciu, 74, 76, chicken
poudre, 12, dust
pouere, 6, 26 ; poure, 134, poor
pouerte, 2, poverty
poun, 74, 76, peacock
pour, 104, fear. B. s.v. *paor*
pourement, 142, poorly
poy, 20 etc., little
poynanz, 20, sharp, keen
pre, 26, 66, 68, 84, 90, 110 ; pree, 102, 104, meadow
prefere, 132, to prefer, like better. C.
premire, 84 ; premere, 22 ; primer, 20, 60, 76, 122, 124, first
prendre, 62 ; prenge, 60 etc.; prent, 37 ; pernez, 34 ; pernent, 34, to take. B. II. 195
presenz, 126, presents
preste, 4, 134, 140, ready
prestement, 124, 132, 134, quickly, easily
prestent, 90 ; prester, to lend
priez, 90, 102, boon-tenants
prime, 6, 36, 37, 60, first
primer, *see* premire
pris, 12, 30 ; prise, 6, 12, 20, 92, 108, price, value, estimation. C.
priue, 104 ; priuetz, 130, familiar, confidential friend. B.
priuement, 130, secretly
prodeshommes, 126, 128 ; prudome, 138, proved men. D. s.v. *prudhomius*
profit, 76, 88, profit
prou, 22, 32, 60 ; pru, 30, 88, 94, 96, 98, 104, 106, 124, 140 ; preu, 36, 88, 102 ; preou, 34, profit
proue, 140, proof
prouendre, 24, 142, provender
prouost, 6, 10, 16, 24, 26, 32, 34, 60, 62, 64, 66, 84, 88, 92, 96, 98, 100, 102, 106, 110, 116, 130, provost
pruant, 88, profitable
pruement, 90, 140, improvement
pur, 2 etc., for
pur, 26, pure
purceaus, 76 ; purceus, 114 ; purceaus letanz, 74 ; purcels, 28, pigs
purceler, 74 ; purcele, 78 ; purcelez, 76, to farrow
purchacer, 122, purchase
purchaz, 64, purchases
pariture, 114 ; purture, 37, disease, rot
purryst, 80 ; puriz, 96 ; porriront, 96 ; purrir etc., to rot, perish
pursuiant, 94, conformable, corresponding to
purueance, 102 ; purueaunce, 124 ; purueiance, 104 ; purueyance, 2, 4, providence, foresight. C.
puruer, 86 ; purueer, 88 ; purveir, 96 ; purveyt, 2, to provide
pus, 2 etc.; puis, 36 etc.; puy, 6, then
pyre, 32 ; pire, 96 ; piur, 96, the worst. B. s.v. *pis*.

Q

qridaunce, 132, credence
quaraunte, 142, forty
quaremel, 6, 8, 14, Lent
quarre, 12, square
quart, 4 etc., fourth
quarter, 66, quarter
quatortzyme, 134, fourteenth
quei, 104, 106 ; quey, 4, why, wherefore
queil, 116 ; quel, 92, why, what
quer, 134, heart
queu, 6, 62, 64, 94, what
quidra, 86 ; quideray, 124 ; quydent, 34 ; quyt, 24 ; quider, to think. B. s.v. *cuider*
quillir, 62 ; quilliz, 96, 102 ; quille, 92 ; quillez, 80, 100 ; cuilla, 134, to gather
quinseyne, 18 ; quinzeyne, 30, fortnight
quint, 16, 68, 74, fifth
quir, 22, 24, 64 ; quyr, 24, skin
quisine, 128, 138, kitchen
qy, 122, 126, who

R

rebiner, 12 ; rebyner, 8, 20, 22 ; rebyne, 12 ; rebynet, 10, 12, to plough the land a second time.
reboutent, 30 ; rebouter, to push back. B.
receueur, 130, receiver
receyte, 32, 126 ; resseit, 60, 64, receipt

GLOSSARIAL INDEX 167

recreaument, 4, grudgingly
recreu, 12; recruz, 22, tired out. C.
refreydyst, 20; refreydyr, to cool
regain, *see* rewain
regete, 22; regeter, to throw back
regne, 84, country. B.
reimsailles, 98, remainder. Cf. R. s.v. *remasilles*
releuer, 22, 32; releuet, 28; releuent, 30, to raise anew, replace. L. s.v. *relever*
relif, 90, relief
religius, 134, religious
relinquissent, 10; relinquir, to leave. B.
remeignanz, 108; remeignant, 92; remeigne, 98; remenge, 142; remenent, 84; remeindre etc., to remain. B. s.v. *manoir*
ren, 2, 14, 18, 90; rien, 60, 62, 104; rin, 88, 90, 92, 98, 102, thing
renable, 132, reasonable
renon, 10, 116, renown, fame
rente, 84, 100, 106, 108, 122, 124, 128, rent; rentes assises, 60, fixed rental
renumee, 104, famed
reon, 8, 12, 14, 92. 'The reon here mentioned seems to include in its breadth the furrow and its accompanying ridge.'—Riley in *Notes and Queries, Second Series*, viii. 32
reoner, 16, to lay in ridges
reperailler, 110, 114, to repair
repose, 12, rest
repris, 4, 6, 10; reprendre, to blame. B. s.v. *prendre*
reprouer, 2, 4, 24, to speak in proverbs. B. s.v. *prover*
rescet, 8, retreat. M. and S. s.v. *receyt*
reson, 26, 34, 84, 88, 92, 104, 108, 118; resun, 4, 62, 66; resoun, 140; reisoin, 128; reyson, 8, reason
respons, 66, 70; respounse, 70, 72, return
respoundre, 64; respoyne, 18; respoyngne, 64; respoyngnent, 74, to return
resseit, *see* receyte
resseiure, 62; resseiue, 64, 66, to receive
retenyr, 22, 28; retigne, 106, to retain
retrere, 62, to withdraw, discharge

reule, 122, 124, 126, 128, 130, 134, 140; rewele, 100, rule
rewain, 12; rewayn, 26; regain, 112, some kind of cheese. 'There is a cheese called a Irwene [rewene, ed. 1567] cheese, the which if it be well-ordered, does passe all other cheses, none excess taken.'—A. Borde, Reg. fol. I. i. *Babees Book* (E.E.T.S.) p. 201
rey, 8, 122; roy, 92, king
robe, 86, 134, 142, robe
Robert Groseteste, 39, 122, Robert Groseteste
rode, 8, 68, 70, 72, rood
roe, 2, wheel
ronge, 30; rounge, 30; roungent, 24; ronger, to ruminate
rouche, 80, hive
roule, 60, 62, 72, 106, 122, 130, roll
rue, 100, street
rychesse, 2, riches, wealth

S

saac, 94; sak, 94, 98, sack
sabelon, 20; sabelun, 20, sand
sabelous, 16, 20, sandy
sache, *see* sauer
sadid, 47, hardened. *Promp. Parv.* s.v. *saddyn*
sakelet, 98, little sack
sakers, 14; sakerer, to draw, drag. B. s.v. *sac*
sale, 134, 138, 140, hall
saler, 28, 30, 116, to salt
salyne, 26, salt, salt-pit
sank, 94, 104, blood
sanz, 6 etc.; santz, 4; saunz, 12 etc., without
sarcher, 16, *read* sarcler
sauagine, 112, 114, 'plentie of wylde beestes.' P.
sauer, 84; sauers, 64; saune, 20, 24; sachant, 86, 108; sachanz, 106; sachaunz, 122, 126; sache, 62; sachent, 10, 20; sauet, 8; seuent, 2, 32, 68; seussent, 62, to know. B. II. 57
saule, 24, 30, surfeited. B. s.v. *saol*
say, 124, *see* sei
sci, 122, here
seal, 128, seal
secher, 28; secchir, 96; sechyr, 30, to dry
secle, 2, 4, 6, world. P.
seculer, 134, secular
secund, 4, 26, 122, second

168 GLOSSARIAL INDEX

seet, 126, 136; seent, 24; set, 2 etc.; seer, to lie down, settle
sei, 72; sey, 66; say, 124, self
seignur, 84, 88, 92, 100, 102, 104, 106, 108, 114, 124, 126, 136, 140; seygnur, 122; seynur, 32, 34; seyngnur, 60, 62, 64, 66, 70, 76, lord
sein, 98, bosom
sein, 36, 96; seyn, 28, 36; seyens, 32, sound
seint—
  seint Barthelmeu, 36, August 24
  seyt croys, 12, May 3
  seint Ioan, 20, 28; seint Iohan, 96; seynt Ioan, 16; seynt Iohan, 12, June 24
  seyn Iude, 32; seynt iude, 56, October 28 (S. Simon and S. Jude)
  seynt luc, 12, October 18
  seint martin, 36, 96, November 11
  seint michel, 26, 37, 62, 76, 96, 126, 142; seynt michel, 18; seint michil, 112; seint mychel, 26; seynt mychel, 18; seint mychil, 118, September 29
  seynt symon, 32, 56 (see seyn Iude)
sek, 14, 28, 76; sekke, 37; siche, 114, dry
sekeresse, 20; sekison, 24, drought
semail, 66, 72, 142; semayl, 8, 12, 16, 18, 22, 'seed sown; also a crop or plants come up of seed; also a sowing; also sowing or seeding time.' C.
semayne, 8, 24, 26; semeyne, 12, 120; semeyngne, 72, 76; simayne, 110, week
semblaunt, 140, manner
semence, 14, 18, 84, 86, 92, 98, 102, 142, seed
semer, 10, 14, 16, 62, 66, 84, 110; seme, 66, 70, 142; semez, 92, to sow
sen, 90, 106, 108, 140, knowledge
seneschal, 62, 86, 88, 90, 92, 104, 106, 122, 124, 126, 130, 138, seneschal
seneschaucie, 84, the office of seneschal
senestre, 14, 34, 138, left
sercler, 18, 68, to weed. L. s.v. sarcler
serement, 122, oath
serianz, 10, 26, 30, 32, 34, 62, 74, 86, 92, 98, 140; seriaunz, 64,
130, 134, 138, 140; serganz, 102, 124, 126; sergaunt, 128, 130, servants
sermenes, 110; surmeynent, 90; surmener, to overdrive
sertefie, 8; sertifye, 6, certified
seruage, 122, service
seruice, 6, 10, 128; seruise, 84, 136, 138, service
seruitor, 138, servant
seson, 12, 14, 16, 20, 22, 28, 32, 94; sesun, 30, 76; seisun, 142; seisoun, 142; seysun, 128, 130; ceson, 22, season
sestzime, 134, sixteenth
setime, 16, 26; setyme, 130, seventeen
seue, 16, moisture
seuerison, 112, weaning
seurement, 118, surely
seussent, see sauer
seute, 86, suit
seyluns, 18; seylloyns, 14, 'a ridge of land lying between two furrows.' H.
seynt Botulf, 144, Boston; seynt yue, 144, S. Ives
si . . non, 62, except. K.
si, 60, here
siches, see sek
sien, 66, 70; sin, 104; soen, 122, his own
sier, 68, 86, 90; sie, 68; siez, 86, 96; syer, 18, to reap
simayne, see semayne
sime, 12, 70; syme, 16, 130, sixth
sin, see sien
sirre, 124; syre, 2, sir
sis, 16, six
soen, see sien
soffrir, 88, 100, 116; sofre, 92; soffert, 128; soffrez, 214; suffrey, 2, to suffer
soil, 142, soil, ground
soiourns, 142; soriorn, 142, sojourn
soleient, 108; soleir, to be accustomed to. B. s.v. soloir
solement, 114, solely
solom, 2 etc.; solum, 4 etc.; sulom, 62 etc., according to. B. s.v. long
somage, 110, burden. K.
somersete, 144, Somerset
somet, 2, summit
soper, 134, 138, 140; super, 138, supper
sor, 98, upon
soruer, see suruer
soruewes, 136, overseers
sostenance, 112; sustenance, 20, 86, 100, sustenance

GLOSSARIAL INDEX 169

soudes, 102, 124, 128, stipend, money. B. s.v. *sol*
souerein, 106, 108, 112; souereyn, 122; soreyne, 84; sourein, 86, 88, sovereign, chief
soule, 98, shoe
soulletz, 134, soiled
soun, *see* sun
sourueyne, *see* suruer
soutz, 2, 26, a shilling. K.
stalun, 64, stallion
strager, 114, to strew
straue, 4, straw
sudeynement, 28, suddenly
suffisante, 24, sufficient
suffisnntment, 22, sufficiently
sufflure, 24, breath
sun, 60, 66, 72; soun, 126, his
surcarke, 86, 100; surcharke, 88; surcarke, 88, 110, overcharge
surdeies, 100, for suzdeis, under-dairymaids
surdre, 2; surd, 132; surse, 32, to rise, spring up
sure, 20, 30, above
surmeynent, *see* sermenes
surplus, 2, 64, 66, 80, 100, surplus
surplusage, 62, surplus
surte, 32, surety
suruer, 10, 92; soruer, 90; sourueye, 10, to oversee
sus, 28, up
sustenance, *see* sostenance
sustenir, 8, to sustain
sustret, 124; sustrete, 84, withheld, drawn. K.
suthampton, 144, Southampton
suzdaye, 116, under-dairywoman
syer, *see* sier
syre, *see* sirre
sywy, 8, followed. B. s.v. *sevre*

T

tabartz, 134, tabard. M. and S.
table, 102, 104, 134, 136, table
tailer, 66; taile, 62, 72, 130; taille, 108, to tally
tart e tempre, 102, late and early. B. II. 330
tas, 72, 92; taas, 126, stack
tascur, 98, 108, a stacker. Cf. 'tascator, a stackman, the maker of a stack, generally of corn.' Raine, *Priory of Finchale* (Surtees Soc.).
tauerne, 100, 114, tavern
tauntost, 62, presently

taylyr, 16, to tally
temoyngnaunce, 62, witness
tempeste, 30, tempest
tempre, 102, early. B. II. 330
tenemenz, 2, 6, 122, tenement, holding
tenet, 2; tent, 6; teigne, 102; tignent, 126; tenir, to hold
tens, 8, 14, 16, 20, 22, 24, 28, 30, 32, 37, 64, 74, 98, 102, 114, time, weather
tenser, 134, to taunt, chide. C.
tenson, 138, dispute
tenur, 110, keeper
terce, 6, 26, 98; terz, 10; tertz, 4; tierce, 124; tierz, 18; tirce, 96; al ters, 10; al tierz, 18, three times, a third
teste, 30, 86, 98; tete, 28, head
testmoyner, 32; testmoyne, 24; tesmoygne, 104, to witness
teus, 132, such
teye, 96, 'any filme or thinne skin.' C.
tiel, 66, such
tignent, *see* tenet
to, 4, two
tollet, 24; tollir, to take away
tondeson, 94; tondeison, 114, shearing
tonel, 78, cask or tun
tor, 106, 112, bull. B.
torcaz, 24, wisp of straw
torner, 68, to turn
torsenouse, 126, wrongful
tost, 96, soon
tot, 6 etc.; toz, 90 etc., all
toteueirs, 74; toteueis, 138, always. B. II. 293
touche, 94. This seems to refer to auncel weight, which was prohibited by Statute 25 Ed. III. c. ix. Cf. H. s.v. *auncel*
toyson, 94, 114, fleece
tramail, 92, spring seed
trames, 66, 70; trameys, 84, corn which ripens within three months of sowing. D. s.v. *tramesium*
transglutent, 30; transglouter, 'greedily to swallow downe his meat halfe chawed.' C.
treche, 114, hurdle
tregst, 86, trickery
treiller, 114; trelee, 118, ' to compasse with latticed frames.' C.
tremutz, 136, feared
trenchet, 16; trencher, to cut
treon, 24; treoun, 30, teat
trere, 96, 112; trete, 114, 128; treet, 2, 100, 116; treez, 96,

112; tret, 96, to milk, draw, etc.
tres, 12, 96; treys, 6 etc., three
treslor, 104, juggler; jogelur and dansur are other readings. Cf. D. s.v. *tresoler*
trespas, 88, 86, 90, trespass
trestoz, 84; trestoutz, 122, 124, 134; trestuz, 126, whole, entire. B. s.v. *tot*
trestzyme, 134, thirteenth
treteyne, 34, *read* certeyne
tricherie, 104, deceit, fraud. B.
troie, 74, 78; troi, 78; troye, 76, 114; treie, 78; true, 28; truye, 28, sow
turberie, 66, turbary

U

uente, 26; vente, 72, 142, sale
uenter, *see* venter
uer, 84 etc.; ueet, 2, to see
ueritable, 104, true
uertetz, 128, warned
ues, 126, your
uespre, 20, evening
ueysin, 4, neighbour
uie, 2, life
uit, 16, eight
uoler, 136, to wish
usage, 122, use, custom
usscher, 134, usher
utime, 16, eighth
unastroille, 134, an idle fellow. *Glossary to West Worcestershire Words* (English Dialect Society)
uus, 142, you

V

v, 142, *read* ou
vache, 22, 24, 26, 74, 76, 78, 86, 92, 94, 96, 108, 112, 118, 142; vacche, 124; vasche, 26, cow
vacher, 112; vachir, 112; wacher, 112, cowherd
vacherie, 112, cowhouse
vadle, 136, valet
valer, 4; valent, 2; vaudra, 2; vaudrunt, 4; vausist, 26, to be worth, useful to. B. s.v. *valoir*
value, 60, 64, 74, value
van, 98, winnowing sieve. C.
veal, 74, 112; veaul, 142; vel, 24, calf
veele, 24, calved

venanz, 102, comers
vente, *see* uente
venter, 74, 76, 130; uenter, 118, to winnow
venteresse, 98, a winnower. D. s.v. *ventrix*
venture, 98, winnowing
venyr, 20; vendra, 14; vendrunt, 16; venent, 62; vennues, 84, 88; veigne, 92; veignent, 36; veyne, 14, 20; vient, 37; vignent, 122 etc.; vynent, 30, to come
ver, 62, 66, 72, 90, 96; veer, 10, 37; veent, 10; veet, 68; veez, 30; veie, 72; vet, 78; veye, 20, to see
vere, 28, boar. B. s.v. *verrat*
verge, 62, 66, 68, 70, rod
vermaille, 36, vermilion
verme, 36, worm. C.
vespre, 36, evening
vessel, 136; vesselle, 116, vessel
vester, *see* voster
vesture, 134, clothing
veue, 32; vewe, 32, view
veylesse, 2, old age
viande, 74; viaunde, 64, meat
viel, 12, 72; ville, 96, 112; veuz, 22, 96, 134, old
vife, 124; viue, 74, alive
vignent, *see* venyr
vigour, 142, vigour
vilain, 106; vilayn, 86, 122; vileyne, 138, bad
vile, 64, 96, 130, township
vilemen, 28, sorely, vilely
villenage, 66, villenage
vin, 128, 138, 142, wine
vint, 12, 128; vintyme, 126, 138; vintesime, 142, twentieth
vinz, 84, a score
viuers, 88, an enclosure in which live game or fish were kept, usually the latter
vnkes, 104, ever. B. s.v. *onkes*
vntzime, 132, eleventh
voisent, *see* aler
volantrif, 104, wilful. B. s.v. *voloir*
volent, 2; voleyent, 66; voyl, 30; voleyr etc., to wish. B.
volente, 132, 134; volunte, 4, will
volenters, 132; volunters, 78, willingly
voster, 28 etc.; vester, 132 etc., your
voye, 8, way
voyr, 16, 22; voyre, 14; voir, 37 veyr, 18, 26, true. B.

GLOSSARIAL INDEX 171

vre, 90, hour
vtime, 16; vtyme, 130; vtre, 134, eighth
vynent, see venyr
vyt, 16, eight

W

wacher, see vacher
wae, 78, a wey
waez, 24, bathed. R.
Walter de henle, 2, Walter of Henley
warenne, 88, warren
waret, 6, 8, 12, 20, 26, 84, 142, land ploughed after a crop, and then left fallow
wareter, 10; waretter, 12, to plough and leave fallow
wascel, 20; wasseus, 72, marsh, wet place. G. s.v. *gacel*. 'Muckhylles,' Lambarde's translation
wast, 6, 26, 88, 140; gast, 16, waste

wastent, see gaste
wauz, 68, waste. Cf. *Waschum*, D.
wedris, 56, wethers
wo, 4, who
wroge, 4: *read* wronge
ws, 2, you. B.
Wyncestre, 144, Winchester
wynyng, 41, gain. M. and S.
wyschous, 4, vicious
wytel, 4, whittle, blanket

Y

yeff, 43, give. M. and S.
yeftis, 43, gifts. M. and S.
yerde, 41, yard of land: a quantity which varied in different localities
yl, 12, he
yrrous, see irous
yssir, see issir
yuer, see iuer
yuernage, see iuernage
yuernail, 84, winter-sowing

PRINTED BY
SPOTTISWOODE AND CO., NEW-STREET SQUARE
LONDON

www.ingramcontent.com/pod-product-compliance
Lightning Source LLC
Chambersburg PA
CBHW031816220426
43662CB00007B/669